34151632

Embryogenesis

the generation of a plant

ENVIRONMENTAL PLANT BIOLOGY series

Editor: W.J. Davies
Institute of Environmental and Biological Sciences, Division of Biological Sciences, University of Lancaster, Lancaster LA1 4YQ, UK

Abscisic Acid: physiology and biochemistry

Carbon Partitioning: within and between organisms

Pests and Pathogens: plant responses to foliar attack

Water Deficits: plant responses from cell to community

Photoinhibition of Photosynthesis: from molecular mechanisms to the field

Environment and Plant Metabolism: flexibility and acclimation

Embryogenesis: the generation of a plant

Forthcoming titles include:

Plant Cuticles: an integrated functional approach

T. L. WANG
Department of Applied Genetics, John Innes Centre, Norwich, UK

A. CUMING
Department of Genetics, University of Leeds, Leeds, UK

Embryogenesis

the generation of a plant

βIOS
SCIENTIFIC
PUBLISHERS

First published 1996

ISBN 1 85996 065 0

BIOS Scientific Publishers Ltd
9 Newtec Place, Magdalen Road, Oxford OX4 1RE, UK
Tel. +44 (0)1865 726286. Fax +44 (0)1865 246823

DISTRIBUTORS

Australia and New Zealand
DA Information Services
   648 Whitehorse Road, Mitcham
Victoria 3132

India
Viva Books Private Limited
   4325/3 Ansari Road, Daryaganj
New Delhi 110002

Singapore and South East Asia
Toppan Company (S) PTE Ltd
   38 Liu Fang Road, Jurong
Singapore 2262

USA and Canada
Books International Inc
   PO Box 605, Herndon, VA 22070

Typeset by Euroset, Alresford, Hampshire, UK
Printed by Information Press, Oxford, UK

Contents

Contributors

Akam, M. Wellcome/CRC Institute, Tennis Court Road, Cambridge CB2 1QR, UK

Alexander, R. Max Planck Institute für Züchtungsforschung, Carl von Linnéweg 10, D-50829, Köln, Germany

Bartels, D. Max Planck Institute für Züchtungsforschung, Carl von Linnéweg 10, D-50829, Köln, Germany

Brown, R.C. Department of Biology, University of Southwestern Louisiana, Lafayette, Louisiana 70504-2451, USA

Butler, W. Department of Genetics, University of Leeds, Leeds LS2 9JT, UK

Clark, J. Department of Biology, University of North Dakota, Grand Forks, ND 58203, USA

Cuming, A.C. Department of Gentics, University of Leeds, Leeds LS2 9JT, UK

Decroocq-Ferrant, V. CSIRO Division of Plant Industry, PO Box 1600, Canberra, ACT 2601, Australia

Doan, D. Plant Molecular Biology Laboratory, Department of Biotechnological Sciences, Agricultural University of Norway, PO Box 5051, N-1432 AAS, Norway

Fischer T. Institut und Lehrstuhl für Genetik, Technische Universität München, Lichtenbergstrasse 4, D-85747 Garching, Germany

Haberer, G. Institut und Lehrstuhl für Genetik, Technische Universität München, Lichtenbergstrasse 4, D-85747 Garching, FRG, Germany

Hedley C.L. John Innes Centre, Colney, Norwich NR4 7UH, UK

Horne, K.L. Department of Botany, University of Leicester, University Road, Leicester LE1 7RH, UK

Johnson, S. John Innes Centre, Colney, Norwich NR4 7UH, UK

Jouannic, S. Université de Paris-Sud, IBP, URA/CNRS 1128, Biologie du Developpement des Plantes, Bâtiment 630, F-91405, Orsay Cedex, France

Kreis, M. Université de Paris-Sud, IBP, URA/CNRS 1128, Biologie du Developpement des Plantes, Bâtiment 630, F-91405, Orsay Cedex, France

Lemmon, B. Department of Biology, University of Southwestern Louisiana, Lafayette, Louisiana 70504-2451, USA

Lindsey, K. Department of Botany, University of Leicester, University Road, Leicester LE1 7RH, UK

Linnestad, C. Plant Molecular Biology Laboratory, Department of Biotechnological Sciences, Agricultural University of Norway, PO Box 5051, N-1432 AAS, Norway

Liu, C.-M. John Innes Centre, Colney, Norwich NR4 7UH, UK

May, V.J. Department of Botany, University of Leicester, University Road, Leicester LE1 7RH, UK

McCarty, D.R. Program of Plant Molecular and Cellular Biology, Horticultural Sciences Department, University of Florida, Gainesville, FL 32611, USA

Meinke, D.W. Department of Botany, Oklahoma State University, Stillwater, OK 74078, USA

Muskett P.R. Department of Botany, University of Leicester, University Road, Leicester LE1 7RH, UK

Olsen, O.-A. Plant Molecular Biology Laboratory, Department of Biotechnological Sciences, Agricultural University of Norway, PO Box 5051, N-1432 AAS, Norway

Raghavan, V. Department of Botany, Ohio State University, 1735 Neil Avenue, Columbus, OH 43210, USA

da Rocha, P.S.C.F. Department of Botany, University of Leicester, University Road, Leicester LE1 7RH, UK

Roncarati, R. Institut für Genetik, Universität Köln, Weyertal 121, D-50931, Köln, Germany

Toonen, M.A.J. Department of Molecular Biology, Agricultural University of Wageningen, Dreijenlaan 3, 6703 HA Wageningen, The Netherlands

Topping, J.F. Department of Botany, University of Leicester, University Road, Leicester LE1 7RH, UK

Torres-Ruiz, R.A. Institut und Lehrstuhl für Genetik, Technische Universität München, Lichtenbergstrasse 4, D-85747 Garching, Germany

Tregear, J. Université de Paris-Sud, IBP, URA/CNRS 1128, Biologie du Developpement des Plantes, Bâtiment 630, F-91405, Orsay Cedex, France

Türet, M. Biology Department, Haccethepe University, Ankora, Turkey

de Vries, S.C. Department of Molecular Biology, Agricultural University of Wageningen, Dreijenlaan 3, 6703 HA Wageningen, The Netherlands

Wang, T.L. Department of Applied Genetics, John Innes Centre, Colney, Norwich NR4 7UH, UK

Wei, W. Department of Botany, University of Leicester, University Road, Leicester LE1 7RH, UK

Abbreviations

2,4-D	2,4 dichlorophenoxyacetic acid	GUS	β-glucuronidase
2iP	N^6-(2-isopentenyl)adenine	IPCR	inverse PCR
BAP	benzylaminopurine	LEA protein	late embryogenesis abundant protein
ABA	abscisic acid	MAP	mitogen-activated protein
ABRE	ABA responsive element	monocot	monocotyledonous plant
Ac	Activator	MOPS	N-morpholine-propanesulphonic acid
AGP	arabinogalactan protein		
AR-h	aldose reductase homologue	MS medium	Murashige–Skoog medium
		NAA	naphthalene acetic acid
CAP	cleaved amplified polymorphism	NCD	nuclear-cytoplasmic domains
		PCR	polymerase chain reaction
DAP	days after pollination	PPB	preprophase band (of microtubules)
dicot	dicotyledonous plant		
dpa	days post-antuesis	RAPD	random amplified polymorphic DNA
Ds	Dissociator		
EGF	epidermal growth factor	RDH	ribitol dehydrogenase
EMS	ethyl methane sulphonate	RFLP	restriction fragment length polymorphism
ERK	external-signal-regulated protein kinase		
		RH	relative humidity
EST	expressed sequence tag	RT-PCR	reverse transcriptase PCR
GA	gibberellic acid	X-gluc	5-bromo-4-chloro-3-indolyl-β-D-glucuronide
GDH	glucose dehydrogenase		
		YAC	yeast artificial chromosome

List of genes and/or gene symbols

N.B. The abbreviation and its explanation are given as cited in the text by the author. They are presented here in capitals as they appear in the index but, they do not necessarily appear in the text in the same manner since each species has its own convention.

AB	*ABORTED SEED*	*AX92*	
ABI1			*BASHFUL*
ABI3		*BIO*	*BIOTIN*
ABI5		*BZ*	*BRONZE*
	ACCESSORY BLADE	*C-MOS*	
	ALBINO	*C1*	
AMP1		*CAT1*	
AP64-13		*COP*	*CONSTITUTIVE PHOTOMORPHOGENIC*
ATMPK1			
ATS1		*CTR1*	

CYD	CYTOKINESIS-DEFECTIVE	
DET	DE-ETIOLATED	
DEK	DEFECTIVE KERNEL	
DEX	DEFECTIVE KERNEL EXPRESSING XENIA	
DEK	DEFECTIVE KERNEL	
EMB	EMBRYO DEFECTIVE	
EMB	EMBRYO-SPECIFIC	
EMP	EMPTY PERICARP	
ENO	ENANO	
EN	ENGRAILED	
FK	FACKEL	
FS	FASS	
	FLACCA	
FUS	FUSCA	
GM	GERMLESS	
	GLABROUSI	
GLB1		
GN	GNOM	
	GOLFTEE	
GPS1		
GP	GROWING POINT	
GSK-3		
GK	GURKE	
GUS		
	HYDRA	
KE	KEULE	
KN	KNOLLE	
KNF	KNOPF	
LAT	LATERNE	
LC		
LEC	LEAFY COTYLEDON	
MCM3		
	MICKEY	
	MIXTA	
MLG3		
MP	MONOPTEROS	
MYB		
MYC		
	NOTCH	
NPK1		
NPT-II		
NTF3		
PEP	PEPINO	
PIN1		
PMEK1		
PRO1		
PRL	PROLIFERA	
PRP8		
PSK4		
PSK6		
R1		
R63-11		
RAB		
RAB17		
RAB28		
	RASPBERRY	
REN	REDUCED ENDOSPERM	
	ROOTLESS	
R	RUGOSUS	
SALT		
SEC7		
SGG	SHAGGY	
STM-1	SHOOT MERISTEMLESS-1	
SHR	SHRUNKEN	
SEX	SHRUNKEN KERNEL EXPRESSING XENIA	
	SIAMENSIS	
	SITIENS	
	SLENDER	
	SPINDLY	
STE11		
STE20		
SUS	SUSPENSOR	
	TRANSPARENT TESTA	
TWN	TWIN	
VP	VIVIPAROUS	
ZLL	ZWILLE	
ZLI	ZYGOTIC LETHAL-1	

Preface

Browsing through a dictionary and indulging in semantics can be absorbing, if time consuming. Trying to find the right term to cover the contents of this book was difficult, satisfying our referees and readers will, no doubt, be even more so. Should the term be embryogeny (the processes by which the embryo is formed; origin, cellular pattern and functions of the embryo), should it be embryogenesis (the origin and formation of the embryo and the science thereof), should it be embryology (the science relating to the embryo and its development) or should it be embryography (the description of the embryo)? Whichever term is chosen, studying the development of the embryo is fascinating and pivotal to any understanding of the processes involved in plant development. In many species, the seed represents the blueprint for the next generation — "the acorn encapsulates the oak" (Sir Walter Scott in 'Waverley'). All the structures crucial to the development of the plant plus its early nourishment are present and, thus, in the embryo, the plant is born.

The chapters in this book represent papers presented at the Society for Experimental Biology's meeting at St. Andrews University on April 3–5, 1995. In these texts, morphological descriptions will be found of the development of crucifers (Chapter 1), grasses (Chapter 6) and legumes (Chapter 12) together with details of the major advances in the field. A broad brush has been used to cover the topic, so that some novel concepts can be embraced, including a chapter devoted largely to endosperm development (see Chapter 10). Several chapters have been allotted to *Arabidopsis* (Chapters 2, 3, 4 and 5) with some inevitable overlaps, but we hope we have kept these to the minimum. Nevertheless, it is only fitting that such prominence should be given to an organism that has been as instrumental in advancing our knowledge of plant development, as has *Drosophila* for animal development, a point underlined in the Foreword. It should not be forgotten, however, that not all plants show the developmental patterns of *Arabidopsis*. Many other species, some of which are covered in this book, have their own advantages as organisms in which to investigate embryogenesis. Comparisons with development in *Drosophila* have proved fruitful for studies on many aspects of plant development and perhaps, again, we should look to this organism for insights into some aspects of plant embryo development. One factor that clearly influences the development of the *Drosophila* embryo, but which is only now receiving the attention it is due by plant researchers, is the maternal influence and the role of stored RNAs (see the Foreword and Chapter 1, Section 1.2.1). The zygotes of most plants clearly possess polarity (along the axis between the micropylar and chalazal ends of the egg cell) in that the first division is asymmetric. This polarity could very easily be established through the influence of the mother plant and eventually lead to the root-shoot axis polarity of the embryo and adult.

Let us hope, however, that many mechanisms will be different between plants and animals, for research would be the poorer, if they were not. As is pointed out in the Foreword, it is likely that those mechanisms governing interactions between cells will differ, since intercellular relations are noticeably different in plants and animals, for example cell movements do not occur in plant embryos. Moreover, somatic embryogenesis can occur in plants (see Chapter 11) so, either one can mimic maternal events by such culture, or additional factors must be involved.

The plant embryo has often been considered an inaccessible structure physically, entombed as it is in the fruit. As a consequence, it has been thought to be removed from scientific analysis. This is clearly no longer the case. Molecular and genetic approaches exemplified in many of the chapters of this volume have opened up the embryo, so to speak, so that we have moved from an era where pure description dominated the literature, to one of analysis, whereby these new approaches come into their own. These molecular analyses also tell us that knowing the identity of a gene responsible for a mutant (see Chapter 3) is where the story begins, linking this knowledge with the phenotype is a much longer tale.

We hope that this text will enable the reader to be up-to-date with recent advances in the field and thus be, if we may be indulgent once again and invent our own terminology to finish this Preface, fully 'embryo-wise'.

Trevor Wang and Andy Cuming
*(on behalf of the Plant Development Group
and the Plant Molecular Biology Group of
the Society for Experimental Biology)*

Acknowledgements

The Editors would like to thank the Society for Experimental Biology (SEB) for organizing the St Andrews meeting, and the Plant Development and the Plant Molecular Biology Groups of the SEB for their financial support. They are also especially grateful to Unilever plc for additional financial support.

Foreword

Embryogenesis — kingdoms apart

M. Akam

The metazoa and the land plants present independent experiments in multicellular development. The embryos of both have solved many of the same developmental challenges: establishing polarity, patterning the distribution of tissues and organs, controlling cell proliferation and defining specialized cell types. Many of these processes are currently best understood in *Drosophila* which has therefore become a point of reference for comparison. My purpose here is to highlight some conclusions from *Drosophila* work that may be of interest to plant embryologists.

Axis formation

The *Drosophila* egg is asymmetrical — an elongated spheroid, slightly flattened on one side, with specialized appendages at one end. In normal development, these asymmetries predict both the antero/posterior and dorso/ventral axes of the future embryo. Thus the cues that define the polarity of the embryo must be laid down during oogenesis.

It has been known for some time how these cues are transmitted to the embryo. Four independent systems are involved, each defined by mutations that affect only one aspect of the embryo's pattern (Nüsslein-Volhard *et al.*, 1987). One defines the dorsal/ventral axis. A second system distinguishes the anterior and posterior cells of the embryo from the central (trunk) region, but does not distinguish anterior from posterior. Two other systems specify the distinct behaviour of anterior and posterior cells.

The anterior and posterior systems are mediated by maternal RNAs and proteins localized at the poles of the egg, within the oocyte. The dorsal and terminal systems use a different mechanism: localized signals are provided by the follicle cells — the somatic cells that ensheathe the forming oocyte, and subsequently secrete the egg-shell (St Johnston and Nüsslein-Volhard, 1992). Polarity is generated in these follicle cells by an interaction with the oocyte during oogenesis, but is only transmitted to the embryo after fertilization, through the action of cell-surface receptors that are uniformly distributed on the surface of the cleaving embryo.

The initial study of these patterning systems identified many gene products that are specific for maternal embryonic patterning. Well-known examples include Bicoid, a homeo-domain transcription factor encoded by the maternal RNA localized at the anterior pole of the egg (Driever and Nüsslein-Volhard, 1988), and Dorsal, a Rel-related transcription factor that enters the nucleus only when activated by the ventral signal from the follicle cells (Roth *et al.*, 1989). Genes such as these are relatively easy to identify by mutational analysis. They have clear maternal-effect phenotypes, uncomplicated by zygotic lethality or pleiotropic effects. The rapid characterization of such genes inevitably focused attention on those aspects of the patterning mechanisms that are specialized adaptations of oogenesis.

More recent results remind us that universal cell biological processes are just as important in embryonic pattern formation (González-Reyes *et al.*, 1995). These new studies focus on the earlier phase of oogenesis: how do the dorsal and terminal follicle cells come to be different from others, and how are different RNA populations localized to anterior and posterior poles of the egg? Polarization of the oocyte cytoskeleton appears to be a key process, guiding both the accumulation of RNA/protein complexes at the poles of the egg, and probably also the migration of the oocyte nucleus. A second key process is the ability of the oocyte nucleus, once localized, to signal to overlying follicle cells. This process, mediated by a secreted signal and its receptor, acts to make posterior follicle cells different from anterior ones, and later (when the oocyte nucleus has migrated from a posterior to antero/dorsal position) to make dorsal follicle cells different from ventral ones. It is not yet known how this signal (a protein related to fibroblast growth factor, encoded by the *Gurken* gene) can be restricted to the vicinity of the oocyte nucleus, but this too must depend on aspects of cell structure.

These early phases of oocyte patterning, which are now seen as crucial for the generation of embryonic polarity in *Drosophila*, build on aspects of cell biology which we may expect to be common to all eukaryotes. Plant embryologists may well find that they have homologues for some of the proteins that act at this phase of embryogenesis — proteins that mediate the intracellular localization and transport of RNA, proteins that orient microtubule arrays, or that allow vectorial export of transcripts from the nucleus.

Subsequent patterning serves to stabilize and transmit these early signals. Some of the mechanisms involved at these later stages are probably ancient and near universal, at least to the metazoa. For example, two key components of the posterior patterning system are Nanos and Pumilio, proteins which act together to block the translation of some maternal RNAs in the posterior of the egg (Murata and Wharton, 1995). Related molecules, doing a similar job, have been implicated in the early patterning of a nematode worm and a vertebrate (Evans *et al.*, 1994). Other of these mechanisms may be highly derived — new uses invented for old molecules. The anterior determinant Bicoid is probably a case in point (Akam *et al.*, 1994). It appears to be a rapidly evolving homeo-domain protein, probably derived from a Hox gene. There is no evidence that a Bicoid-like determinant is used to transmit maternal information in lower insects, and certainly no Hox gene with an analogous role in vertebrates.

My own guess is that Dorsal too will prove to be a protein newly recruited to the control of embryonic polarity. Dorsal homologues in insects and vertebrates share an involvement in the control of the immune response (Ip *et al.*, 1993). The *Dorsal* gene of insects may have been derived from such a gene, in a classic case of duplication and functional divergence. (The functional divergence does not appear to be complete, for although Dorsal is not required for immune function in *Drosophila*, it is still induced in the appropriate tissue when the immune system is activated.)

Embryogenesis

The earliest stages of patterning in the *Drosophila* embryo appear to be highly specialized. Intracellular gradients of transcription factors are set up in the cytoplasm of the egg during cleavage stages, before membranes have partitioned the nuclei into separate cells (Driever and Nüsslein-Volhard, 1988). The first of these morphogens is encoded by maternal RNAs. Subsequent rounds of patterning are mediated by the products of the gap and pair-rule segmentation genes, which divide the embryo into finer and finer domains of gene activity (Pankratz and Jackle, 1993). Most of the genes used in this patterning process are also used during other, well-conserved stages of embryogenesis, for example in the nervous system. However, the regulatory interactions that characterize the segmentation gene hierarchy probably represent a novel mechanism that evolved within insects to exploit the possibilities offered by syncytial cleavage (Akam and Dawes, 1992). Their study has provided a fine opportunity to dissect the integration of information by complex promoters (e.g. Stanojevic *et al.*, 1991), but we do not yet know how to relate this information to the early patterning of other embryos.

The opposite is true of later embryogenesis, after the embryo has become cellular. Most of the mechanisms operating after gastrulation appear to have parallels in vertebrates, and utilize recognizably homologous molecules. Among the most striking examples are the use of the secreted proteins Hedgehog and TGFβ-related molecules to mediate embryonic induction (Fietz *et al.*, 1994; Hogan *et al.*, 1994), the use of Hox genes to specify position along the antero-posterior axis (McGinnis and Krumlauf, 1992), and the use of trans-membrane proteins of the Notch/Delta family to mediate lateral inhibition between adjacent cells (Chitnis *et al.*, 1995). These parallels go beyond the common inheritance of basic cell biological processes, and presumably reflect the inheritance and conservation of multicellular patterning mechanisms from a common ancestor that was itself multicellular. Sticking my neck out, I would hazard a guess that such intercellular mechanisms will not be shared by plants and animals, in contrast to the intracellular signalling cascades, mechanisms for cytoskeletal control, and the like. It will be fascinating to see how the plant and animal kingdoms have taken the same basic 'eukaryote tool kit' and elaborated it to solve, in their own fashion, the problems of multicellular development.

References

Akam, M. and Dawes, R. (1992) More than one way to slice an egg. *Curr. Biol.* **2**, 395–398.

Akam, M., Averof, M., Castelli-Gair, J., Dawes, R., Falciani, F. and Ferrier, D. (1994) The evolving role of Hox genes in arthropods. *Development*, 1944 (Suppl), 209–215.

Chitnis, A., Henrique, D., Lewis, J., Ish-Horowicz, D. and Kintner, C. (1995) Primary neurogenesis in *Xenopus* embryos regulated by a homologue of the *Drosophila* neurogenic gene *Delta*. *Nature* **375**, 761–766.

Driever, W. and Nüsslein-Volhard, C. (1988) The bicoid protein determines position in the *Drosophila* embryo in a concentration-dependent manner. *Cell* **54**, 95–104.

Evans, T. C., Crittenden, S. L., Kodoyianni, V. and Kimble, J. (1994) Translational control of maternal glp-1 messenger RNA establishes an asymmetry in the *C. elegans* embryo. *Cell* **77**, 183–194.

Fietz, M. J., Concordet, J.-P., Barbosa, R., Johnson, R., Krauss, S., McMahon, A. P., Tabin, C. and Ingham, P. W. (1994) The *hedgehog* gene family in *Drosophila* and vertebrate development. *Development*, 1994 (Suppl), 43–51.

González-Reyes, A., Elliott, H. and St Johnston, D. (1995) Polarization of both major body axes in *Drosophila* by *gurken-torpedo* signalling. *Nature* **375**, 654–658.

Hogan, B. L. M., Blessing, M., Winnier, G. E., Suzuki, N. and Jones, C. M. (1994) Growth factors in development: the role of TGFβ-related polypeptide signalling molecules in embryogenesis. *Development*, 1994 (Suppl), 53–60.

Ip, Y. T., Reach, M., Engstrom, Y., Kadalayil, L., Cai, H., González-Crespo, S., Tatei, K. and Levine, M. (1993) *Dif*, a *dorsal*-related gene that mediates an immune response in *Drosophila*. *Cell* **75**, 753–763.

McGinnis, W. and Krumlauf, R. (1992) Homeobox genes and axial patterning. *Cell* **68**, 283–302.

Murata, Y. and Wharton, R. (1995) Binding of Pumilio to maternal *hunchback* mRNA is required for posterior patterning in *Drosophila* embryos. *Cell* **80**, 747–756.

Nüsslein-Volhard, C., Frohnhofer, H. G. and Lehmann, R. (1987) Determination of antero-posterior polarity in *Drosophila*. *Science* **238**, 1675–1681.

Pankratz, M. J. and Jackle, H. (1993) Blastoderm segmentation. In: *The Development of* Drosophila melanogaster (eds M. Bate and A. Martinez-Arias). Cold Spring Harbor, New York, pp. 467–516.

Roth, S., Stein, D. and Nüsslein-Volhard, C. (1989) A gradient of nuclear localization of the dorsal protein determines dorsoventral pattern in the *Drosophila* embryo. *Cell* **59**, 1189–1202.

St Johnston, D. and Nüsslein-Volhard, C. (1992) The origin of pattern and polarity in the *Drosophila* embryo. *Cell* **68**, 201–220.

Stanojevic, D., Small, S. and Levine, M. (1991) Regulation of a segmentation stripe by overlapping activators and repressors in the *Drosophila* embryo. *Science* **254**, 1385–1387.

Perspectives on the molecular cytology of embryogenesis

V. Raghavan

1.1 Introduction

The logical starting point for the study of embryogenesis is the zygote, which is the first cell of the sporophytic generation. This cell is endowed with all the genetic information necessary to construct the adult organism in full multi-cellularity, sexuality and structure, and it does so by progressive divisions to form the embryo. In angiosperms, development of the embryo involves extensive changes in form accompanied by the fabrication of new tissues and organs at the same time as the participating cells undergo a progressive change from an undifferentiated to a differentiated state. These events occur in the embryo which is housed in a privileged location in the embryo sac that is itself buried within the tissues of the ovule. Presumably, during this critical stage, the developing embryo is nourished by the maternal tissues.

The study of embryogenesis in angiosperms has a long history, dating back to the Greek philosophers. Early studies up to about 1870 were chiefly concerned with the concept of sexuality in plants, and were devoted to an understanding of the basic structural features of the male and female reproductive parts of the flower. A fundamental contribution to the study of embryogenesis was made by Hanstein in 1870 with the publication of his work on the development of embryos of certain dicotyledons and monocotyledons. The choice of *Capsella bursa-pastoris* (Brassicaceae) by this investigator was indeed a fortunate one, for in subsequent years this species has become a textbook example of early embryo-genesis in a dicot. It is to Hanstein's credit that he correctly identified the position of walls during successive divisions of the zygote; this investigator introduced such terms as the quadrant and octant to designate the four-celled and eight-celled stages, respectively, of the globular embryo, and the term hypophysis to denote to the cell at the lower end of the globular embryo. These terms are to this day in the literature. Following the publication of Hanstein's work and other works of similar nature, a new insight into the reproductive biology of angiosperms emerged from the discovery of syngamy in plants by Strasburger in 1877 and the discovery of the more unique phenomenon of double fertilization independently by Nawaschin and Guignard in 1899.

During the period from about 1910 to 1950 the emphasis was placed on the descriptive and comparative aspects of embryo development in angiosperms. Investigations undertaken on microsporogenesis, megasporogenesis, embryogenesis and endosperm development in a wide range of plants provided a useful framework for an appreciation of the variation in the development and organization of the male and female reproductive units of angiosperms as well as the variation in the process of embryogenesis. During this period when descriptive embryogenesis flourished, advances made in the fields of plant physiology, biochemistry and genetics and refinements in tissue culture techniques have had a substantial impact on the direction of embryological research. This gave rise to the discipline of experimental embryogenesis (1950–1970), involving the culture of flowers, ovaries, ovules and embryos in order to study embryogenesis under controlled conditions and the culture of isolated tissues, organs, single cells, protoplasts, anthers and pollen grains to deflect their normal developmental programme in the embryogenic pathway. Thus, experimental embryogenesis transformed an era of purely descriptive studies into one of experiments and deductions, designed to determine the physiological basis for the precisely programmed development of the embryo within the confines of the ovule.

With regard to complexity, it is perhaps one further step from experimental embryogenesis to an area in the study of angiosperm embryogenesis which has been designated as molecular embryogenesis. In general, studies in this area are concerned with the spatial and temporal activation of genes involved in the development of the embryo from its single-celled beginning, isolation and cloning of genes, determination of the amino acid sequence of the proteins encoded by the genes and identification of the protein products. In recent years, research in this field has made dramatic progress, leading to an increasing awareness of the real potential of genetic engineering for improvement of our agricultural crops.

Investigations of molecular embryogenesis have utilized a combination of many different methods. Because of the availability of large quantities of embryos at different stages of maturity, genes that perform critical roles during the maturation phase of embryos have been identified and characterized in a number of plants using standard molecular techniques. Most importantly, these include genes for storage protein synthesis, desiccation tolerance, defence-related functions and vivipary or precocious germination in the seed. However, such methods cannot be used to study the molecular biology of fertilization or the pattern of gene expression during the division of the zygote and proembryos, nor are these methods suitable for monitoring spatial patterns of gene expression in specific parts of the mature embryo such as the shoot apical meristem, which is composed of no more than a few thousand cells. The major obstacle to these studies is the fact that the events of fertilization and embryogenesis occur within the relatively inaccessible environment of the embryo sac and it is impossible to collect sufficient material of the fertilized egg or the early division phase embryos for traditional molecular studies. Of the two strategies widely used in the analysis of early development of the embryo, one that has proved to be very useful is the isolation and characterization of developmental mutants. The other strategy has been to monitor the accumulation of nucleic acids and other macromolecules cytochemically, to determine their synthesis by autoradiographic methods, and to localize bulk mRNA and specific transcripts by *in situ* hybridization with cloned genes and examine their expression by tagging to reporter genes for histochemical detection. The application of these methods to the study of angiosperm embryo developmental biology collectively constitutes the basis of the molecular cytology of embryogenesis.

The purpose of this chapter is to outline how the study of embryo development in angiosperms can be approached from a molecular cytological perspective. The treatment given is mainly in the form of an overview of the landmark changes identified during embryogenesis using the techniques of molecular cytology, rather than an attempt to present a thorough review of the methods and results obtained. This is followed by a discussion of the results of an investigation of the localization of mRNA in developing embryos of *C. bursa-pastoris*.

1.2 An overview of the molecular cytology of embryogenesis

Molecular cytological methods have been used to study a broad range of problems in embryogenesis, such as the search for stored mRNAs, constancy of DNA in the cells, changes in RNA and proteins foreshadowing critical divisions of the zygote and proembryos and spatial localization of mRNAs and cloned genes during embryogenesis. These investigations have obvious implications for gene activity during embryogenesis, beginning with the zygote, and progressing through the developing embryo stages to the mature embryo.

1.2.1 *Search for stored mRNAs*

Very little is known about the molecular biology of fertilization and the mechanism of gene activation during the early division of the zygote in angiosperms, nor are the relative contributions of the maternal genome and the paternal genome to the division of the zygote well understood. One of the established concepts in animal embryology is that the egg contains a considerable store of mRNA which becomes available after fertilization to code for the first proteins of the zygote. There is no hard evidence for the existence of a legacy of template information in the angiosperm egg, and the source of the mRNA parcelled out for the division of the zygote is unknown. An investigation by Nagato (1979) has addressed this problem by culturing young florets of *Oryza sativa* (rice) and *Hordeum vulgare* (barley) in media containing precursors of RNA ([^{3}H]-uridine) or protein ([^{3}H]-leucine) synthesis. It was found that, although no autoradiographically detectable incorporation of [^{3}H]-uridine occurs in developing embryos composed of less than 100 cells, [^{3}H]-leucine is incorporated into embryos of all ages. Since the eggs are fertilized after the florets are placed in culture, it has been argued that protein synthesis in the absence of concurrent RNA synthesis probably involves the use of mRNA stored in the egg and activated after fertilization. Although changes in the size of the precursor pools cannot be ruled out under the experimental conditions employed, in general the results support the thesis that *de novo* transcription is limited at least up to the globular stage of embryo development.

1.2.2 *Constancy of DNA in the cells*

The concept of constancy of DNA in a cell predicts that DNA synthesis must keep pace with the production of cells so that each new cell will have a constant amount of DNA (known as the C value). One might expect that the precise division of the zygote to produce a multicellular embryo will depend upon an efficient DNA duplication mechanism followed by cytokinesis. This picture is obscured in some studies because of the complications introduced by endoreduplication. The DNA content of the zygote examined by Feulgen microspectrophotometry has been shown to range from the expected 2C amount to 4C amounts in *Tradescantia paludosa* (Woodard, 1956), *Hordeum vulgare* and *H. vulgare* × *H. bulbosum* (Bennett and Smith, 1976), and to reach 3C to 6C amounts in *Petunia hybrida* (Vallade *et al.*, 1978) or become as high as 16C in *H. distichum* (Mericle and Mericle, 1970). The DNA contents of nuclei of two- to four-celled proembryos of these species show a similar range, from 2C in *T. paludosa* to 8C in *H. distichum*. In the globular embryo of *Vanda sanderiana* a gradient of increasing DNA values from 2C to more than 8C can be seen in cells from the distal meristematic tip to the proximal parenchymatous region (Alvarez, 1968). As has been observed in the torpedo-shaped embryo of *Vitis vinifera*, with further development the DNA content of the cells stabilizes at the 2C level (Faure and Nougarède, 1993). Analyses of embryos of *Lactuca sativa* (Brunori and D'Amato, 1967) and *Vicia faba* (Brunori, 1967) have shown that the DNA content of cells does not deviate from the 2C level even as dehydration preparatory to quiescence of the seed sets in. It is possible that, with progressive desiccation of the embryo, other factors such as available water supply might limit DNA synthesis.

Much of our present knowledge of the dynamics of DNA synthesis and accumulation in developing embryos has been derived from analyses of cotyledons of legumes. After an initial period of growth by cell division, subsequent growth of the cotyledons is characterized by cell expansion accompanied by the accumulation of an acervate complex of storage products, mostly proteins. Scharpe and van Parijs (1973) showed that cells of developing cotyledons of *Pisum sativum* (pea) continue to synthesize DNA after cell divisions have stopped, leading to a progressive increase in the DNA content of cells of fully grown cotyledons to a level as high as 64C (see also Chapter 12, Liu *et al.*). The general pattern of changes in the DNA levels of cotyledonary cells of *P. arvense* closely resembles that described in *P. sativum* (Smith, 1973). In the cotyledons of

Vicia faba (Millerd and Whitfeld, 1973) and *Gossypium hirsutum* (cotton) (Walbot and Dure, 1976) it has been established that the increase in DNA content of the cells during the phase of cell elongation occurs by endoreduplication. The fact that endoreduplication occurs at the same time as accumulation of storage proteins is of interest with regard to the possible relationship between the two processes.

Investigations of the nuclear cytology of suspensor cells of various angiosperms have shown that their development is characterized by endoreduplication coupled with the presence of polytene chromosomes. A recurrent theme emerging from these studies is that, despite its short life span, the suspensor is composed of transcriptionally active cells which may play a regulatory role in the development of the embryo. On the basis of nuclear volume or quantitative microspectro-photometry, values for the DNA content of suspensor cells reported to date range from 16C in *Sophora flavescens* (Nagl, 1978) to 8192C in *Phaseolus coccineus* (Brady, 1973). In *P. coccineus* (Brady, 1973; Nagl, 1974) and *Eruca sativa* (Corsi *et al.*, 1973), the outcome of endoreduplication in cells in the different regions of the suspensor is somewhat different, but its consequences in terms of gene regulation are probably the same. In both these species there is a gradient of DNA content from low values in the cells of the suspensor close to the embryo, to intermediate values in the neck region and very high values in the cells confined to the basal region. Endoreduplication in the suspensors of *Alisma lanceolatum* and *A. plantago-aquatica* is unconventional, since it affects only the large basal cell. The nucleus of this cell rhythmically bloats in size and, on the basis of the increase in nuclear volume, it has been estimated that in the latter species the nucleus attains a ploidy level as high as 1024n (Bohdanowicz, 1973, 1987). Although the chromosomal constitution of endoreduplicated suspensor cells of several species has revealed polyteny, the phenomenon has been most extensively studied in *P. coccineus* (Nagl, 1967) and *P. vulgaris* (Nagl, 1969a, b). Like polytene chromosomes in the salivary glands of dipteran larvae, those of *P. coccineus* and *P. vulgaris* display puffs which represent sites of DNA synthesis and transcription. A gene identified in the polytene chromosome of *P. coccineus* by *in situ* hybridization using a cloned probe has been shown to encode a protein that inhibits fungal endopolygalacturonase (Frediani *et al.*, 1993).

As has been shown by molecular hybridization of 25S, 18S and 5S [³H]-rRNA fractions to DNA of suspensor cells of *P. coccineus*, genes for rRNA are clustered at the nucleolus-organizing regions of polytene chromosomes (Durante *et al.*, 1977). However, only the ribosomal cistrons embedded in random nucleolus-organizing centres are functionally active and synthesize RNA in a given chromosome (Frediani *et al.*, 1986). In general, these bands engaged in RNA puffing are undermethylated. Consequently, the transcriptional activity of polytene chromosomes may rely on specific molecular components of DNA. When the activity of polytene chromosomes at different stages of embryogenesis is analysed by [³H]-uridine incorporation, these chromosomes arc found to engage in RNA synthesis in different ways (Forino *et al.*, 1992). Another intriguing observation is that differing labelling patterns appear in the polycistronic transcription units for 25S, 18S and 5S rRNA genes when polytene chromosomes are annealed with fractions of DNA banded in the analytical ultracentrifuge. This shows that rRNA cistrons of the same class are interspersed with different DNA sequences (Durante *et al.*, 1987). These observations are illuminating and provide a good foundation for the goal of achieving a complete understanding of the transcriptional apparatus of suspensor polytene chromosomes.

The question as to whether polytene chromosomes undergo selective DNA amplification has been subject to intensive study at the cytological level. Auto-radiography of [^{3}H]-thymidine incorporation into suspensor cells of *P. coccineus* showed that, in most cells, the label is found exclusively in the heterochromatic regions of the polytene chromosomes, while in a small number of cells the label is found throughout the nucleus. These results have been interpreted as indicating that certain genes of the heterochromatic regions undergo DNA amplification in addition to the scheduled DNA synthesis due to endoreduplication (Avanzi *et al.*, 1970). Consistent with the observation that cells in the different regions of the suspensor display varying degrees of endoreduplication, selective DNA amplifi-cation is found to prevail in the puffs of the highly polytene chromosomes of cells in the micropylar region of the suspensor, while cells produced during the early stages of suspensor development show uniform labelling throughout the nuclei due to endoreduplication (Cremonini and Cionini, 1977). While these observations highlight a regulatory role for the polytene chromosomes of the suspensor cells, precise determination of the function of their gene products, given the small size and the transient nature of the suspensor, requires a protracted study.

1.2.3 *Changing trends in RNA and protein metabolism during early embryogenesis*

Once fertilization has been accomplished, many properties of the egg change during the first few hours of its existence as a zygote. At the ultrastructural level, these changes consist mainly of establishment of polarity by rearrangement of organelles and increase in the number and complexity of the organelles, indicating of increased metabolic activity of the cell. Particularly significant is the observation that the generation of new ribosomes and the formation of polysomes from existing ribosomes represent part of the basic response of the egg to fertilization (Raghavan, 1976, 1986). Consistent with the conclusion, based on these observations, that fertilization signals the production of mRNA and its engagement by ribosomes, it has been shown that in *Capsella bursa-pastoris* (Schulz and Jensen, 1968) and *Nicotiana rustica* (Sehgal and Gifford, 1979), fertilization is accompanied by an increase in the stainable RNA content of the zygote. *In situ* hybridization of ovules of *C. bursa-pastoris* with [^{3}H]-polyuridylic acid ([^{3}H]-poly(U)) has revealed not only an increase in the mRNA content of the zygote, but also a gradient in its accumulation, with a high concentration in the chalazal part (Raghavan, 1990).

In most angiosperms, the first division of the zygote is asymmetrical and transverse, cutting off a large vacuolate basal cell toward the micropylar end and a small densely cytoplasmic terminal cell. With regard to the fate of these cells, the terminal cell gives rise to the organogenetic part of the embryo, with or without contributions from the derivatives of the basal cell, which forms a filamentous suspensor. Histochemical studies have shown that the marked degree of structural and functional asymmetry displayed by the two-celled proembryo is complemented by subtle variations in the distribution of RNA. In *Gossypium hirsutum* (Jensen, 1963) and *Vanda* (Alvarez and Sagawa, 1965), the difference between the terminal and basal cells is signalled by a high concentration of RNA in the former. By contrast, in *Stellaria media* (Pritchard, 1964), *C. bursa-pastoris* (Schulz and Jensen, 1968) and *Limnophyton obtusifolium* (Shah and Pandey, 1978), the RNA concentration is higher in the basal cell than in the terminal cell. In the early-stage embryos of other plants that have been examined, a characteristic RNA distribution pattern is the presence of higher levels of stainable RNA in the cells of the presumptive cotyledons than in the cells of the rest of the embryo (Norreel, 1972; Rondet, 1962; Syamasundar and Panchaksharappa, 1976; Vallade, 1970). Taken together these observations are not surprising, since it is reasonable to assume that the transcriptional activity of the cell will increase in preparation for differentiation.

Another measure of the functional differentiation of cells of developing embryos is the distribution of proteins and enzymes. In *C. bursa-pastoris*, consistent with the increased concentration of RNA in the zygote and in the basal cell of the two-celled proembryo, these cells are found to stain densely for proteins (Schulz and Jensen, 1968). Evidence has been presented which indicates that in *S. media* there is a difference in the magnitude of acid phosphatase activity between the different cell groups of the heart-shaped embryo and between the embryo and the suspensor. In general, enzyme activity is confined to the cells of the embryo and is absent from the suspensor. Within the embryo itself, the enzyme is present in the protoderm, but is absent from the hypophysis which forms the root cap. Surprisingly, alkaline phosphatase is present throughout the embryo and the suspensor (Pritchard and Bergstresser, 1969). A striking correlation between tissue and organ differentiation and the activity of succinic dehydrogenase, a key enzyme in aerobic respiration, has been demonstrated in developing embryos of cotton. This is manifest in the presence of high enzyme titre in all recognizable organs or parts of the embryo, including the suspensor, cotyledons, hypocotyl, shoot apex, subapical region and the elongating radicle (Forman and Jensen, 1965). These observations not only give meaning to the process of embryogenesis, but also suggest that a cytochemical examination of embryos with more refined techniques should prove instructive.

1.2.4 *Spatial distribution of cloned genes*

By *in situ* hybridization techniques, it is possible to localize cloned genes and nucleic acid molecules in cytological preparations at the level of the tissue or the cell. The attractions of this method for monitoring the accumulation of transcripts of specific genes or bulk mRNA during the early division phases of embryogenesis are obvious. The method uses both radioactive and non-radioactive probes and can localize either DNA or RNA in tissue sections.

Since the axis of the embryo does not become visibly distinct until about the heart-shaped stage, it is of interest to identify the earliest developmental stage when cell specifications occur. One attempt to solve this problem involved *in situ* localization of transcripts of *Glycine max* (soybean) Kunitz trypsin inhibitor gene (KTi3) in developing soybean embryos (Perez-Grau and Goldberg, 1989). This work showed that Kunitz trypsin inhibitor mRNA accumulates specifically at the micropylar end of the globular-stage embryo and then becomes localized within the axis of the heart-shaped, cotyledon-stage and maturation-stage embryos, indicating that the cells destined to become the axis are already specified at the globular stage. Changes in the spatial and temporal distribution of transcripts of the storage proteins, napin and cruciferin, in developing embryos of *Brassica napus* have revealed patterns that reflect early determinative events related to the establishment of boundaries set up by the first transverse division of the proembryo. It is deduced that the patterns of accumulation of transcripts do not reflect differences in tissue or cell type, on the basis of the observation that cell derivatives from the lower tier of the proembryo cells accumulate high levels of storage protein messages in the torpedo-shaped embryo. In these embryos, only cotyledons derived from the upper tier accumulate the messages (Fernandez *et al.*, 1991).

Genes that exhibit highly regulated expression in embryos have been introduced into transgenic plants in order to identify *cis* control regions that regulate their expression. The tissue- or cell-specific pattern of expression of the inserted gene is generally followed by analysis of β-glucuronidase (GUS) activity in the embryos of transgenic plants by a simple histochemical assay. A gene designated *AX92* is expressed in the torpedo-shaped embryos of *Brassica napus*, but it is not detected in late-stage embryos. Transgenic technology has shown that the expression of this gene is limited to the root cortical cells of torpedo-shaped embryos, and that a combination of *cis*-acting DNA sequences in the 5′ and 3′ flanking and/or untranslated regions influences the expression of this gene (Dietrich *et al.*, 1992). In a recent review, Goldberg *et al.* (1994) cite unpublished experiments showing that, whereas the KTi3/GUS fusion gene is expressed in the micropylar end of globular embryos of transgenic *Nicotiana tabacum* (tobacco), expression of a chimeric lectin/GUS gene is confined to the equatorial region of such embryos. This has led to the conclusion that discrete transcriptional domains are established in the apical–basal and radial axes of the globular embryo as a prelude to differentiation of specific regions at later stages of embryogenesis.

In addition to the traditional embryonic organs, embryos of members of the Poaceae (e.g. *Oryza sativa*) are composed of additional organs such as the coleoptile, coleorhiza, epiblast, scutellum and mesocotyl. We followed the localization of a rice histone H3 gene and a rice glutelin gene in developing embryos of rice in order to determine whether these organs arise from cell types which can be identified at an early stage of embryogenesis (Raghavan and Olmedilla, 1989; Ramachandran and Raghavan, 1990). The first signs of accumulation of histone and glutelin messages in developing embryos are observed at the club-shaped stage. During subsequent development, pronounced localization of histone mRNA is seen in the cells of the coleoptile and the first leaf, but the shoot apex and the leaf primordia lack the label. Histone mRNA is expressed in all parts of the root primordium, including the coleorhiza, the root cap and the zone of elongation, but signals are not detected in the quiescent centre. The pattern of binding of the histone probe in the different parts of the embryo remained essentially unchanged during the 20-day developmental period studied, but there is an appreciable decay of the histone message during grain maturation. Maximum annealing of glutelin gene transcripts is found in embryos which have initiated shoot and root meristems; subsequent growth of embryos is accompanied by a loss of binding sites, which completely disappear even before the embryo matures. Although glutelin gene expression is initiated simultaneously in the embryo and the endosperm, accumulation of the message in the endosperm continues for several days after the message has peaked in the embryo.

Accumulation of histone gene transcripts in the non-dividing cells of the embryo, such as the coleoptile, leaf and elongating zone of the root, is somewhat unusual since it is well-known that histones complex with DNA during cell division. At a general level, the pattern of histone gene accumulation revealed by these studies is one of absence of transcripts during the early stages of embryo-genesis, their appearance during mid-developmental stages, and their decay with maturity of the grain. As has been shown by Northern blot analysis, analogous changes occur in the prevalence of cloned storage protein genes during embryogenesis in soybean (Goldberg *et al.*, 1981). The data for glutelin gene expression show that a certain degree of morphological differentiation is required before the embryo begins to accumulate these gene transcripts. The early decay of glutelin mRNA in the embryo is probably due to its limited capacity as a storage organ, compared to the endosperm.

The localization of specific genes in embryos of normal and transgenic plants is usefully considered to be a significant improvement over simple histochemical and cytochemical techniques for following determinative events of early embryo-genesis and eventually identifying the genes responsible for controlling the basic architecture and morphogenesis of embryos.

1.2.5 *Localization of mRNA*

Since most mRNA is tailed by a polyadenylic acid [poly(A)] segment, it is possible to localize mRNA in cytological preparations by using [^{3}H]-poly(U) as a probe. Annealing sites due to binding of the probe are detected by autoradiography. *In situ* hybridization of sections of developing embryos of *Capsella bursa-pastoris* has enabled us to provide evidence for the single-celled origin of the quiescent centre in an angiosperm root, and to suggest a role for the hypophysis in such a root (Raghavan, 1990). As in other angiosperms, the first division of the zygote of *C. bursa-pastoris* is transverse and yields a large vacuolate basal cell toward the micropylar end of the ovule and a small densely cytoplasmic terminal cell toward the chalazal end. Subsequently, the basal cell divides transversely to produce a three-celled proembryo. Following *in situ* hybridization, autoradiographic silver grains resulting from [³H]-poly(U) binding are concentrated at the chalazal end of the zygote. After the first division of the zygote, the silver-grain density in the terminal and basal cells remained more or less unchanged. No differences were observed in the density of labelling between the cells of the three-celled proembryo, indicating that a renewed accumulation of poly(A) does not occur in the cells of the proembryo.

In *C. bursa-pastoris*, the middle cell of the three-celled proembryo gives rise, by a series of transverse divisions, to an eight- or nine-celled filamentous suspensor, subtended at the micropylar end by the large basal cell. As the suspensor is formed, only background levels of radioactivity are detected in its cells. The suspensor cell closest to the organogenetic part of the crucifer type of embryo is the hypophysis. In the embryo of *C. bursa-pastoris*, the hypophysis divides transversely to form two cells; in the mature embryo, the root cortex is derived by further divisions of the inner cell, whereas the root cap and root epidermis are contributed by the descendants of the outer cell (Schaffner, 1906). There is an interesting distribution of label within the cells of the root apex, beginning with the globular embryo. At this stage of embryogenesis, autoradiographic silver grains appear to be localized in all cells of the presumptive root apex, except for those in the hypophysis, while during later stages of embryogenesis, daughter cells cut off by the hypophysis also lack the label. Cytological examination showed that progressive divisions of the inner cell of the hypophysis contribute to the formation of a hemispherical group of three tiers of cells in the radicle of the mature embryo. This group of cells, which occupies the position of the quiescent centre in the root apex, is bounded on the inside, toward the embryo axis, by the procambium, and on the outside by the outer cell of the hypophysis and its derivatives. In contrast to the other labelled cells of the root apex of the mature embryo, cells of the quiescent centre characteristically lack [³H]-poly(U) binding sites. Presumably, the lack of poly(A) + RNA in the cells of the quiescent centre is due to the origin of the latter from a progenitor cell which itself is not enriched with mRNA sequences.

In order to confirm the identity of the quiescent centre and to define it in metabolic terms, a comparison was made between autoradiographs of sections of seedling roots of *C. bursa-pastoris* subjected to *in situ* hybridization with [³H]-poly(U) and those allowed to incorporate [³H]-thymidine. It was observed that, during the first 2 days of germination, meristematic activity results in the formation of three or four layers of cells at the root apex constituting the root cap and a layer of cells of the root epidermis. Moreover, the same three tiers of cells of the root apex delimited by the procambium on the inside and by the root epidermis on the outside remain unlabelled by both methods, showing that they are composed of the specific cells of the quiescent centre. At the metabolic level, the quiescent centre has long been characterized by its low DNA synthetic activity (Clowes, 1961). RNA polymerase activity and the rates of RNA and protein synthesis are also lower in the cells of the quiescent centre than in the surrounding cells (Clowes, 1956, 1958; Fisher, 1968; Jensen, 1958). The results described here provide yet another negative criterion for identification of the cells of the quiescent centre, namely that they do not serve as a site of mRNA accumulation.

1.3 Conclusions

Our current knowledge of embryogenesis in angiosperms owes its origin to foundations laid in the past, especially to the descriptive accounts of embryo development in a vast array of plants. Due to the dramatic progress made in the field of molecular biology, there has been a radical reorientation in our outlook on this subject. I consider the molecular cytological approach to the study of embryogenesis to represent the middle ground between the old and the new, yet to be essential for defining embryo development in all its complexity, beginning with the single-celled zygote. What has emerged from the review presented in this chapter is that the molecular organization of developing embryos as studied by cytological techniques is exceedingly complex, and a great deal of research with early division phase embryos is still needed to investigate the mechanisms whereby the limited population of cells interact and communicate with each other to evolve into the mature embryo.

References

Alvarez, M.R. (1968) Quantitative changes in nuclear DNA accompanying postgermination embryonic development in *Vanda* (Orchidaceae). *Am. J. Bot.* **55**, 1036–1041.

Alvarez, M.R. and Sagawa, Y. (1965) A histochemical study of embryo development in *Vanda* (Orchidaceae). *Caryologia* **18**, 251–261.

Avanzi, S., Cionini, P.G. and D'Amato, F. (1970) Cytochemical and autoradiographic analyses on the embryo suspensor cells of *Phaseolus coccineus. Caryologia* **23**, 605–638.

Bennett, M.D. and Smith, J.B. (1976) The nuclear DNA content of the egg, the zygote and young proembryo cells in *Hordeum. Caryologia* **29**, 435–446.

Bohdanowicz, J. (1973) Karyological anatomy of the suspensor in *Alisma* L. I. *Alisma plantago-aquatica* L. *Acta Biol. Craco. Ser. Bot.* **16**, 235–246.

Bohdanowicz, J. (1987) *Alisma* embryogenesis: the development and ultrastructure of the suspensor. *Protoplasma* **137**, 71–83.

Brady, T. (1973) Feulgen cytophotometric determination of the DNA content of the embryo proper and suspensor cells of *Phaseolus coccineus. Cell Differentiation* **2**, 65–75.

Brunori, A. (1967) Relationship between DNA synthesis and water content during ripening of *Vicia faba* seed. *Caryologia* **20**, 333–338.

Brunori, A. and D'Amato, F. (1967) The DNA content of nuclei in the embryo of dry seeds of *Pinus pinea* and *Lactuca sativa. Caryologia* **20**, 153–161.

Clowes, F.A.L. (1956) Nucleic acids in root apical meristems of *Zea. New Phytol.* **55**, 29–34.

Clowes, F.A.L. (1958) Protein synthesis in root meristems. *J. Exp. Bot.* **9**, 229–238.

Clowes, F.A.L. (1961) *Apical Meristems.* Blackwell Scientific Publications, Oxford.

Corsi, G., Renzoni, G.C. and Viegi, L. (1973) A DNA cytophotometric investigation on the suspensor of *Eruca sativa* Miller. *Caryologia* **26**, 531–540.

Cremonini, R. and Cionini, P.G. (1977). Extra DNA synthesis in embryo suspensor cells of *Phaseolus coccineus. Protoplasma* **91**, 303–313.

Dietrich, R.A., Radke, S.E. and Harada, J.J. (1992) Downstream DNA sequences are required to activate a gene expressed in the root cortex of embryos and seedlings. *Plant Cell* **4**, 1371-1382.

Durante, M., Cionini, P.G., Avanzi, S., Cremonini, R. and D'Amato, F. (1977) Cytological localization of the genes for the four classes of ribosomal RNA (25S, 18S, 5.8S and 5S) in polytene chromosomes of *Phaseolus coccineus*. *Chromosoma* **60**, 269-282.

Durante, M., Cremonini, P.G., Tagliasacchi, A.M., Forino, L.M.C. and Cionini, P.G. (1987) Characterization and chromosomal localization of fast renaturing and satellite DNA sequences in *Phaseolus coccineus*. *Protoplasma* **137**, 100-108.

Faure, O. and Nougarède, A. (1993) Nuclear DNA content of somatic and zygotic embryos of *Vitis vinifera* cv. Grenache noir at the torpedo stage. Flow cytometry and *in situ* DNA microspectrophotometry. *Protoplasma* **176**, 145-150.

Fernandez, D.E., Turner, F.R. and Crouch, M.L. (1991) *In situ* localization of storage protein mRNAs in developing meristems of *Brassica napus* embryos. *Development* **111**, 299-313.

Fisher, D.B. (1968) Localization of endogenous RNA polymerase activity in frozen sections of plant tissues. *J. Cell Biol.* **39**, 745-749.

Forino, L.M.C., Tagliasacchi, A.M., Cavallini, A., Cionini, G., Giraldi, E. and Cionini, P.G. (1992) RNA synthesis in the embryo suspensor of *Phaseolus coccineus* at two stages of embryogenesis, and the effect of supplied gibberellic acid. *Protoplasma* **167**, 152-158.

Forman, M. and Jensen, W.A. (1965) Respiration and embryogenesis in cotton. *Plant Physiol.* **40**, 765-769.

Frediani, M., Forino, L.M.C., Tagliasacchi, A.M., Cionini, P.G., Durante, M. and Avanzi, S. (1986) Functional heterogeneity, during early embryogenesis, of *Phaseolus coccineus* ribosomal cistrons in polytene chromosomes of embryo suspensor. *Protoplasma* **132**, 51-57.

Frediani, M., Cremonini, R., Salvi, G., Caprari, C., Desiderio, A., D'Ovidio, R., Cervone, F. and De Lorenzo, G. (1993) Cytological localization of the *PGIP* genes in the embryo suspensor cells of *Phaseolus vulgaris* L. *Theor. Appl. Genet.* **87**, 369-373.

Goldberg, R.B., Hoschek, G., Ditta, G.S. and Breidenbach, R.W. (1981). Developmental regulation of cloned superabundant embryo mRNAs in soybean. *Dev. Biol.* **83**, 218-231.

Goldberg, R.B., de Paiva, G. and Yadegari, R. (1994) Plant embryogenesis: zygote to seed. *Science* **266**, 605-614.

Jensen, W.A. (1958) The nucleic acid and protein content of root tip cells of *Vicia faba* and *Allium cepa*. *Exp. Cell Res.* **14**, 575-583.

Jensen, W.A. (1963) Cell development during plant embryogenesis. *Brookhaven Symp. Biol.* **16**, 179-202.

Mericle, L.W. and Mericle, R.P. (1970) Nuclear DNA complement in young proembryos of barley. *Mutat. Res.* **10**, 515-518.

Millerd, A. and Whitfeld, P.R. (1973) Deoxyribonucleic acid and ribonucleic acid synthesis during the cell expansion phase of cotyledon development in *Vicia faba* L. *Plant Physiol.* **51**, 1005-1010.

Nagato, Y. (1979) Incorporation of [^{3}H]uridine and [^{3}H]leucine during early embryogenesis of rice and barley in caryopsis culture. *Plant Cell Physiol.* **20**, 765-773.

Nagl, W. (1967) Die Riesenchromosomen von *Phaseolus coccineus* L.: Baueigentümlichkeiten, Strukturmodifikationen, zusätzliche Nukleolen und Vergleich mit den mitotischen Chromosomen. *Österr. Bot. Z.* **114**, 171-182.

Nagl, W. (1969a) Banded polytene chromosomes in the legume *Phaseolus vulgaris*. *Nature* **221**, 70-71.

Nagl, W. (1969b) Puffing of polytene chromosomes in a plant (*Phaseolus vulgaris*). *Naturwissenschaften* **56**, 221-222.

Nagl, W. (1974) The *Phaseolus* suspensor and its polytene chromosomes. *Z. Pflanzenphysiol.* **73**, 1-44.

Nagl, W. (1978) *Endopolyploidy and Polyteny in Differentiation and Evolution*. North-Holland Publishing Co., Amsterdam.

Norreel, B. (1972) Etude comparative de la répartition des acides ribonucléiques au cours de l'embryogenèse zygotique et de l'embryogenèse androgénétique chez le *Nicotiana tabacum* L. *Compt. Rend. Acad. Sci. Paris* **275D**, 1219-1220.

Perez-Grau, L. and Goldberg, R.B. (1989) Soybean seed protein genes are regulated spatially during embryogenesis. *Plant Cell* **1**, 1095-1109.

Pritchard, H.N. (1964) A cytochemical study of embryo sac development in *Stellaria media. Am. J. Bot.* **51**, 371-378.

Pritchard, H.N. and Bergstresser, K.A. (1969). The cytochemistry of some enzyme activities in *Stellaria media* embryos. *Experientia* **25**, 1116-1117.

Raghavan, V. (1976) *Experimental Embryogenesis in Vascular Plants*. Academic Press, London.

Raghavan, V. (1986) *Embryogenesis in Angiosperms. A Developmental and Experimental Study*. Cambridge University Press, Cambridge.

Raghavan, V. (1990) Origin of the quiescent center in the root of *Capsella bursa-pastoris* (L.) Medik. *Planta* **181**, 62-70.

Raghavan, V. and Olmedilla, A. (1989) Spatial patterns of histone mRNA expression during grain development and germination in rice. *Cell Differ. Dev.* **27**, 183-196.

Ramachandran, C. and Raghavan, V. (1990) Intracellular localization of glutelin mRNA during grain development in rice. *J. Exp. Bot.* **41**, 393-399.

Rondet, P. (1962) L'organogenèse au cours de l'embryogenèse chez l'*Alyssum maritimum* Lamk. *Compt. Rend. Acad. Sci. Paris* **255**, 2278-2280.

Schaffner, M. (1906) The embryology of the Shepherd's purse. *Ohio Naturl.* **7**, 1-8.

Scharpe, A. and van Parijs, R. (1973) The formation of polyploid cells in ripening cotyledons of *Pisum sativum* L. in relation to ribosome and protein synthesis. *J. Exp. Bot.* **24**, 216-222.

Schulz, R. and Jensen, W.A. (1968) *Capsella* embryogenesis: the egg, zygote, and young embryo. *Am. J. Bot.* **55**, 807-819.

Sehgal, C.B. and Gifford, E.M. Jr (1979) Developmental and histochemical studies of the ovules of *Nicotiana rustica* L. *Bot. Gaz.* **140**, 180-188.

Shah, C.K. and Pandey, S.N. (1978) Histochemical studies during embryogenesis in *Limnophyton obtusifolium. Phytomorphology* **28**, 31-42.

Smith, D.L. (1973) Nucleic acid, protein, and starch synthesis in developing cotyledons of *Pisum arvense* L. *Ann. Bot.* **37**, 795-804.

Syamasundar, J. and Panchaksharappa, M.G. (1976) A histochemical study of some post-fertilization developmental stages in *Dipcadi montanum* Dalz. *Cytologia* **41**, 123-130.

Vallade, J. (1970) Développement embryonnaire chez un *Petunia hybrida* hort. *Compt. Rend. Acad. Sci. Paris* **270D**, 1893-1896.

Vallade, J., Cornu, A., Essad, S. and Alabouvette, J. (1978) Niveaux de DNA dans les noyaux zygotiques chez le *Petunia hybrida* hort. *Soc. Bot. Fr. Actual. Bot.* **125**, 253-258.

Walbot, V. and Dure, L. III. (1976) Developmental biochemistry of cottonseed embryogenesis and germination. VII. Characterization of the cotton genome. *J. Mol. Biol.* **101**, 503-536.

Woodard, J.W. (1956) DNA in gametogenesis and embryogeny in *Tradescantia. J. Biophys. Biochem. Cytol.* **2**, 765-776.

2

Genes involved in the elaboration of apical pattern and form in *Arabidopsis thaliana*: genetic and molecular analysis

R.A. Torres-Ruiz, T. Fischer and G. Haberer

2.1 Introduction

The generation of the plant seedling is the result of fundamental processes which are at work from early on in embryogenesis. Pattern formation addresses the organizational aspect of the seedling body. It involves those factors and events which cause cell types, tissues and organs of the plant body to originate at specific locations such that a structurally and functionally meaningful context or pattern is established. The overall shape of the plant body is generated by cell shape changes and by cell division planes, which alternate in orientation and frequency in dividing cells at different locations. (For convenience the term morphogenesis will be used in this article for generation of body shape.) Differentiation is the term used to cover the generation of cellular diversity, while growth refers to the process which increases the size of the developing organism. Obviously these processes are somehow linked to each other and must be precisely coordinated. Thus, a major task for plant developmental biologists will be to identify the responsible factors (genes), their interdependences and their temporal and spatial regulation during embryogenesis.

The adult plant body is generated during two periods of development (Steeves and Sussex, 1989). Following the embryo and seedling stage, the terminally located meristems are responsible for further development and generate the adult structures of the plant. With regard to pattern formation, therefore, the extent to which pattern information from the seedling stage is needed for the organization of the adult plant body, and 'how much' or which information is carried by the meristems, is of considerable interest. An understanding of the relationship between embryo and adult development will also clarify whether pattern processes differ qualitatively between both stages. At least in the case of flowers, several genes have been identified which seem to be solely responsible for the elaboration of these organs (Coen and Meyerowitz, 1991; Schwarz-Sommer et al., 1990).

Developmental work in the animal field has shown that progress in understanding biological problems is critically dependent on the organism chosen for study. For instance, the peculiarities of the *Drosophila* system yielded considerable insight into pattern formation of the fruit fly (Ingham, 1988; Nüsslein-Volhard, 1991; Slack, 1993). Classical plant embryology has provided considerable information about the early cell division patterns of a number of plant species (Johri, 1984; Rutishauser, 1969). However, the complexity of embryo development forces one to concentrate on a few and, in different respects, suitable systems. In monocotyledonous plants (monocots), rice and maize have been preferentially investigated (Kitano et al., 1993; Nagato et al., 1989; Neuffer and Sheridan, 1980; Sheridan and Clark, 1993). In dicotyledonous plants (dicots), descriptive as well as molecular analyses of embryos have been carried out. In soybean, several genes encoding seed storage proteins have been cloned and their expression patterns analysed (e.g. Goldberg et al., 1989; Perez-Grau and Goldberg, 1989). Detailed embryo analysis has been performed for the crucifers *Capsella bursa-pastoris, Brassica napus* and *Arabidopsis thaliana* (Jürgens and Mayer, 1994; Mansfield and Briarty, 1991; Müller, 1963; Schulz and Jensen, 1968; Tykarska, 1976, 1979). However, due to the unique peculiarities of the system, further work, particularly that aiming to identify embryo developmental genes by mutagenesis, has concentrated on *Arabidopsis thaliana* (Feldmann, 1991; Jürgens et al., 1991; Mayer et al., 1991; Meinke, 1985; Meinke and Sussex, 1979; Müller, 1963). These attempts have ultimately led to the recent cloning of the embryo pattern gene *EMB30/GNOM* (Shevell et al., 1994).

2.2 The use of *Arabidopsis thaliana* as a system for studying plant embryogenesis

The crucifer *Arabidopsis thaliana* combines several advantageous features which make it an ideal experimental subject. These include the small size of the plant and its nuclear genome, short generation time, self fertilizing hermaphroditic flowers and the large number of progeny per flower (Meyerowitz, 1989; Redei, 1975). In addition, *Arabidopsis* does not possess a starchy endosperm. This has been particularly advantageous for the analysis of embryo stages in whole-mount preparations of entire immature seeds.

Concentrating on one or a few systems raises the question of whether insights into one system are of general significance. Several points of evidence suggest that such a generalization may be valid. Flowering plants form a taxonomic group of closely related species (Takhtajan, 1980) for which a monophyletic origin is likely (Crane *et al.*, 1995). Although they have quite different adult forms, flowering plants pass (at least initially) through comparable stages of embryo-genesis (Johri, 1984; Rutishauser, 1969; Steeves and Sussex, 1989). The seedling appears to be quite different in dicots and monocots, but essentially has a similar body organization, which can be described as the superimposition of two patterns: the apical-basal axis and the radial axis (Jürgens *et al.*, 1994). The apical-basal axis consists of the epicotyl (including the shoot meristem), cotyledon(s), hypocotyl, embryonic root (radicle), root meristem and root cap. The radial pattern includes the epidermis, ground tissue and central vascular tissue. Further-more, work on homeotic flower genes has shown that in plants as diverse as *Arabidopsis* and *Antirrhinum*, homologous genes perform very similar tasks (Coen and Meyerowitz, 1991; Schwarz-Sommer *et al.*, 1990; Weigel *et al.*, 1992). Hence, it is reasonable to assume that the study of embryogenesis in one system, such as that of *Arabidopsis thaliana,* might well contribute to the understanding of plant embryo development in general.

2.3 *Arabidopsis* embryo development

The body organization of the seedling is laid down during embryogenesis. Essentially two phases of development can be identified. Developmental processes leading to the heart stage establish the primary body plan which can be recognized later in the seedling. This phase represents about 30% of embryo development (Jürgens *et al.*, 1991). During the following phase, the size of the embryo is increased and maturation of the embryo is completed prior to the dormancy stage (Mansfield and Briarty, 1991). Embryo development in *Arabidopsis* follows a uniform and invariable sequence of cell division patterns in the early embryo stages (Figure 2.1). It starts with the unequal division of the zygote, which produces a small, but cytoplasm-rich, apical cell and a larger, vacuolated basal cell (Jürgens and Mayer, 1994; Mansfield and Briarty, 1991; Schulz and Jensen, 1968). This unequal division is anticipated and probably supported by a marked polarization of the egg cell, expressed by a polar distribution of cytoplasm, vacuoles and nucleus (Mansfield *et al.*, 1991). The daughter cells generated by

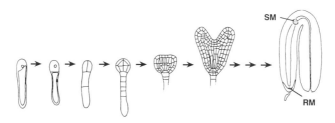

Figure 2.1. *Stages of* Arabidopsis *embryo development (from left to right): zygote, two-cell stage, quadrant stage, 16-cell stage, globular stage, heart stage and torpedo stage. SM = shoot meristem; RM = root meristem.*

this division develop into the embryo proper and the suspensor, respectively, and the latter is thought to be responsible for nourishing the embryo. Interestingly, *Arabidopsis* suspensor mutants show that some genes which are involved in proembryo development also participate in the establishment of the filamentous suspensor (Schwartz *et al.*, 1994). The apical cell undergoes further vertical divisions which lead to the quadrant stage. Then, after a horizontal division, the octant-stage embryo develops, which is organized into two layers of four cells each. These layers constitute the upper and lower tier, respectively. The horizontal boundary between these tiers is still detectable in the more advanced heart stage. The upper tier contributes solely to the apical part of the seedling, with cotyledons and apex (including the shoot meristem precursors), while the lower tier gives rise to the hypocotyl and root. The lower tier might also contribute to the basal parts of the cotyledons as shown by sector analysis (Scheres *et al.*, 1994). The next series of divisions is tangential and establishes the precursor of the epidermis. At this time the upper most cell of the suspensor (the hypophysis) intrudes into the proembryo and contributes to the elaboration of the root primordium, including the meristematic cells, epidermis and lateral root cap initials. During the following division and cell expansion events the globular embryo stage develops, which includes the epidermis precursor, root primordium, ground tissue and elongated vascular cell precursors and two layers of apical cells in the upper part of the proembryo. The radial symmetry of the globular embryo is altered during the next series of divisions. In the upper region these divisions initially lead to a flattening of the surface at the end of the late globular stage and the embryo enters the triangular stage. This latter stage is brief, since ongoing divisions immediately produce the cotyledon primordia which grow above the level of the shoot pole. In order to generate these primordia the epidermis undergoes several anticlinal divisions, the frequency of which is particularly high in the regions where the incipient cotyledon primordia become visible. Underneath the epidermal layer, anti- and periclinal divisions generate two sub-epidermal layers which contribute to the widening of the upper half and the development of the cotyledons (Jürgens and Mayer, 1994; Mansfield and Briarty, 1991). Judging from their location, the presumptive shoot meristem precursor cells reside in the centre of the upper region in between the cotyledon buttresses. However, it is not known whether this group of cells is already committed to develop into the shoot meristem. A morphological distinction between shoot meristem and other cells is only possible in the torpedo stage (Barton and Poethig, 1993). Cell divisions and cell shape changes cause the lower layer to be twice as long as the upper region. By the end of these processes, the embryo has adopted a bilateral symmetry which reflects the organization of the seedling. Accordingly, the following divisions mainly result in growth of the organ primordia, and do not add new structures. In more advanced (torpedo) stages, the frequency of cell divisions declines and eventually stops. From advanced torpedo stages onwards the cells undergo fundamental physiological changes associated with reserve deposition and maturation of the embryo (Mansfield and Briarty, 1992). At the end of these events the embryo enters dormancy and prepares for germination.

2.4 Approaches to the analysis of pattern formation and morphogenesis in flowering plants

Diverse approaches have been adopted in order to analyse the principles of embryo pattern formation and morphogenesis. Classical histological studies have revealed that certain plant families can be characterized by invariable patterns of cell division and cell shape changes (see Johri, 1984, and references therein). Additional information has been obtained using experimental approaches which aimed to investigate the response of somatic embryos after dissection by micro-surgery (Schiavone and Racusen, 1990). These studies showed that different parts of the carrot embryo have different developmental capabilities (e.g. the shoot pole is able to regenerate the root pole fully, but the converse is not always true). Somatic embryos have also been used to test the significance of extracellular glycoproteins in embryo development (de Jong *et al.*, 1993; van Engelen and de Vries, 1992; Chapter 11, Toonen and de Vries). A further peculiarity of somatic embryos is that they pass through similar stages to zygotic embryos, although they exhibit different cell division patterns (Halperin, 1966; McWilliam *et al.*, 1974). Physiological experiments have indicated the involvement of hormones in development, especially in elaboration of the cotyledons (Liu *et al.*, 1993). Molecular analysis of embryogenesis has revealed a number of genes that are expressed during the early and late stages of embryogenesis in soybean and carrot (e.g. Perez-Grau and Goldberg, 1989; Sterk *et al.*, 1991), some of which may be used as molecular markers that indicate the presence or absence of specific cells or cell groups. A different approach, termed 'genetic dissection', utilizes mutants as the starting point for genetic and molecular analysis. A screening strategy to identify pattern genes in *Arabidopsis* has been developed, which was based on the idea that such mutants should proceed through embryogenesis and germinate as abnormal seedlings (Jürgens *et al.*, 1991). Those seedling abnormalities which specifically alter the body organization should be of interest, and the observed abnormalities should be detectable at embryo stages which are critical for elaboration of the respective pattern elements (Jürgens *et al.*, 1991). This strategy led to the isolation of apical-basal and radial pattern mutants (Mayer *et al.*, 1991). These were named *gnom* (*gn*), affecting the apical and basal end (also designated emb30; Meinke, 1985), *gurke* (*gk*), affecting the apical region, *monopteros* (*mp*), affecting the basal region, and *fackel* (*fk*), affecting the central part of the seedling. Mutants *keule* (*keu*) and *knolle* (*kn*) interfere with the radial pattern of the seedling.

2.5 Mutants which interfere with the elaboration of the apical pattern

The key for identifying genes involved in the elaboration of the apical pattern is to search for mutants which exhibit specific defects in the upper half of the seedling, and which become manifest no later than the heart stage (possible exceptions may be mutants which solely affect the shoot apical meristem). Lines from chemical, X-ray, fast-neutron and T-DNA insertional mutagenesis have been screened for such phenotypes, and a number of lines segregating mutants with apical defects have been found (Mayer *et al.*, 1991, 1993a; A. Lohner and R. A. Torres-Ruiz, unpublished data). To date, at least three loci representing such genes have been found (Figure 2.2). Despite the fact that numerous mutants have been found which satisfy the criteria mentioned, the respective genes are represented by quite different numbers of alleles. The gene *GURKE* (*GK*) is represented by at least 15 alleles, while the gene *PEPINO* (*PEP*) is defined by only two alleles. The

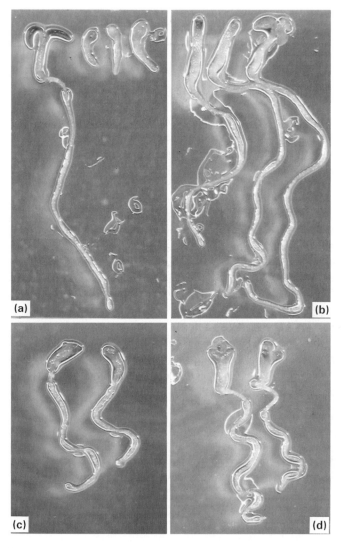

Figure 2.2. Arabidopsis *apical pattern mutants: (a) wild-type and* gk; *(b)* lat; *(c)* pep; *(d)* R63-11.

gene *LATERNE* (*LAT*) is represented by one allele, as are the genes *R63-11* and *AP64-13* (further apical pattern candidates).

Mutations of the *gk*, *PEP* and *R63-11* genes lead to apical defects in which hemispherical rudiments of cotyledons occur in the proper seedling position (Figures 2.2 and 2.3). The root is often shorter than in the wild type, but retains all the elements of the wild-type root, namely root hairs, root meristem and root cap, and all the tissues defining the radial axis. It has yet to be established whether the shortness of the root (especially in strong *gk* mutants) is a direct or indirect effect, but it seems probable that it results from the loss of the main parts of the cotyledons, which normally provide the germinating seedling with basic nutrients. The phenotype is most dramatic in some strong *gk* alleles. The respective lines segregate phenotypes with rudimentary cotyledons but also phenotypes which have no cotyledons at all. In extreme cases most of the hypocotyl region is also absent (R.A. Torres-Ruiz, A. Lohner and G. Jürgens, in preparation). The two mutant alleles of *PEP* are phenotypically indistinguishable from weak *GK* alleles, but complementation analysis shows that *GK* and *PEP* are different genes. Preliminary complementation tests indicate that the line *R63-11* represents a third gene with mutants very similar to those of *PEP* and *GK*. However, in some cases the cotyledons which are initially rudimentary may expand upon further growth, and some individuals develop protuberances on the margins of the

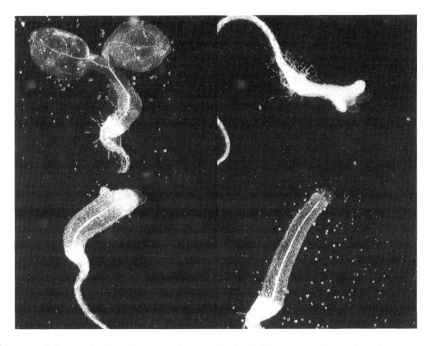

Figure 2.3. *Apical pattern mutants. Dark field preparations showing internal (e.g. vascular) structures: wild-type (upper left);* gk *(upper right);* pep *(lower left);* lat *(lower right corner).*

cotyledon rudiments. In some alleles, leaf-like structures may develop on the tip of the mutant apex. At least in the case of *GK* this does not lead to normal development of the shoot (R.A. Torres Ruiz, A. Lohner and G. Jürgens, personal communication).

In *gk*, *pep* and *R63-11* it is obvious that the hypocotyl is either not fully developed or not elongated along the apical-basal axis as in the wild type. The situtation is completely different in *lat* seedlings, which show no obvious defect in their hypocotyls (Figure 2.3). The striking feature of *lat*, which is so far represented by only one allele, is that in most phenotypes the cotyledons are very precisely deleted (Figures 2.2 and 2.3). Interestingly, tricotyledonous and rarely monocotyledonous seedlings are also consistently segregated in this line (R.A. Torres-Ruiz, unpublished data). Thus it is probable that this mutation leads to variable cotyledon phenotypes. In some cases, further development of *lat* is possible, and leads to adult plants with abnormal stems terminating in abnormal flower structures, terminal leaves, or without any further structure (R.A. Torres-Ruiz, unpublished data).

The phenotypes of the mutants described above suggested that the observed alterations should be visible, at least at the heart stage of the embryo where the cotyledon primordia are established. Indeed, the *GK*, *pep*, *R63-11* and *lat* mutants all show apical defects at this stage (R.A. Torres-Ruiz, A. Lohner and G. Jürgens, unpublished data). It remains to be determined whether (subtle) deviations occur in the earlier stages (e.g. during the globular or early heart stage). Figure 2.4 shows the deviation of an apical defect mutant (*R63-11*). The comparison with wild type embryos at the same stage reveals several features. The observed deviations do not concern the suspensor but only the embryo proper and are mainly localized at the apical end of the embryo. Localized alterations of cell-division planes lead to an apical end which, in contrast to that of the wild-type embryo, does not clearly develop two separate cotyledon primordia. The root primordium and elements of the radial pattern of the hypocotyl appear to be less, if at all, disturbed. During subsequent development the growth of the mutant embryo is retarded. While the cotyledons of the wild type expand significantly, the mutant embryo barely elaborates cotyledons; only rudiments of these structures can be seen in torpedo-stage embryos (Figure 2.4).

On the basis of their phenotypes and stage of deviation from the wild type, apical defect mutants can be separated from several other known embryo pattern mutants (Figure 2.5). In most cases the first manifestation of the mutant phenotype occurs at the stage when the corresponding structure, which is defective in these mutants, is elaborated in the wild type (see Figure 2.5 for overview). For instance, the terminal pattern mutant *emb30/gnom* (*gn*), which shows defects in polarity, has already deviated from normal by the first division of the zygote (Mayer *et al.*, 1993b). Monopteros mutants (*mp*), which lack a root and hypocotyl, are abnormal in the basal region of early globular stages (Berleth and Jürgens, 1993), and *fackel* mutants (*fk*) which are defective in the central part of the body, deviate from normal at the heart stage (Mayer *et al.*, 1991). The same is true for the radial pattern mutants *keule* (*keu*) and *knolle* (*kn*) (Mayer *et al.*, 1991). The shoot apical

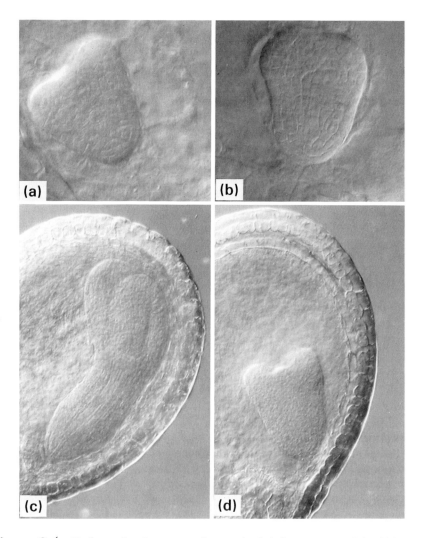

Figure 2.4. *Embryo development of an apical defect mutant: (a) wild-type heart stage; (b) mutant heart stage; (c) wild-type torpedo stage; (d) mutant torpedo stage.*

meristem mutants *zwille* (*zll*) (Jürgens *et al.*, 1994) and *shoot meristemless-1* (*stm-1*) (Barton and Poethig, 1993) deviate at a very late stage.

This raises the question of how the different events are coordinated and integrated. Double-mutant analysis will elucidate some interactions, but is proving to be difficult at the moment. To date it has been demonstrated that *gn* is epistatic to *mp* (Mayer *et al.*, 1993b). Double-mutant analysis of apical pattern gene combinations between *gk* and *pep* and *gk* and *lat* did not reveal clear dependences

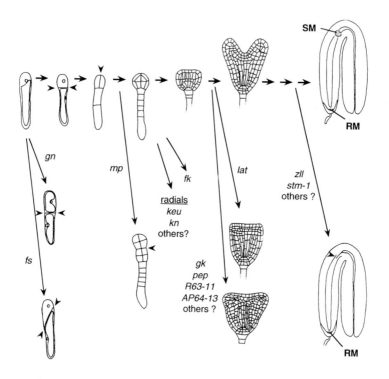

Figure 2.5. *Embryo pattern and form mutants of* Arabidopsis: *stages of their divergence from the wild type (for references see text). Arrows point to stages of divergence and arrowheads to critical cell divisions in the zygote, the quadrant and the octant stage, respectively. SM = shoot meristem, RM = root meristem. (For abbreviations of mutants see text.)*

(unpublished observations). In these cases PCR-assisted analysis with closely linked molecular markers will be essential (see below).

2.6 Mutants which alter the shape of the seedling

The aforementioned mutants have shown that pattern alterations are correlated with localized abnormal orientations of cell-division planes. However, a group of mutants exhibiting a dramatically altered body shape has demonstrated that oriented cell divisions are not instrumental in pattern formation (Mayer *et al.*, 1991; Torres-Ruiz and Jürgens, 1994). Initially, three genes of this class, *FASS* (Figure 2.6), *KNOPF* (Figure 2.6) and *MICKEY*, were found (Mayer *et al.*, 1991).

Another candidate for this group is the *ENANO* gene (Figure 2.6). The common characteristic of these mutants is that their whole body shape is altered in a very

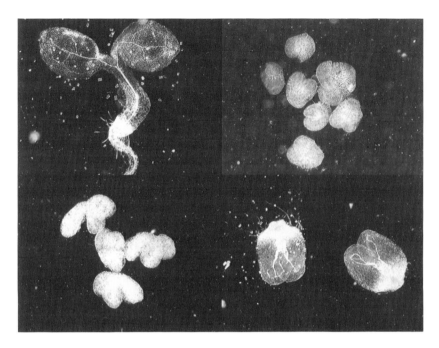

Figure 2.6. Arabidopsis *shape mutants (dark field preparations): wild-type (upper left);* knopf *(upper right);* enano *(lower left);* fass *(lower right corner).*

specific way but all pattern elements originate at their appropiate (wild-type) positions. This phenotypic specificity allowed the identification of several alleles for each gene (Mayer *et al.*, 1991), and distinguishes this group of mutants from those leading to variable and unspecific phenotypes (Jürgens *et al.*, 1991). Despite this common characteristic, the phenotypes of the different shape genes are quite dissimilar. The phenotypes of *MICKEY* and *ENANO* exhibit a significant retardation in growth, but this seems to be even more marked in *fass* (*fs*) and *knopf* (*knf*) seedlings, which display an extreme compression along the axes of the body organs (Mayer *et al.*, 1991; Torres-Ruiz and Jürgens, 1994; R. A. Torres-Ruiz, unpublished data). The retardation in growth is accompanied in *knf* and *enano* (*eno*) mutants by a significant decrease in greening of the cotyledons while *fs* and *mickey* exhibit normal greening. All mutants display alterations in the shape of the cells, the consequence of which is a drastic change in body shape in comparison to the wild type. A detailed analysis of *fs* mutants revealed that all pattern elements originate at their appropriate location, although cell-division planes may deviate markedly from those of the wild type during all stages of embryo development (Torres-Ruiz and Jürgens, 1994). However, changes in cell shape may be quite different between shape mutants. In *fs* and *eno*, cell walls are quite irregularly oriented and lead to the development of cells of different sizes, although they belong to the same tissue (e.g. epidermis). In contrast, cells of *knf* mutants have regularly oriented cell walls which have elongated inappropriately (data not shown). In *fs* embryos the mutation may disturb the orientation of the first cell division in the zygote (Torres-Ruiz and Jürgens, 1994). However, it is not known whether the mutants displaying other shapes diverge at earlier stages than the heart stage.

All observations concerning shape genes suggest that they might well code for quite disparate functions which are directly or indirectly involved in the elaboration of cell shapes. This is not surprising in view of the complex array of factors which influence and establish the cytoskeleton, which is in turn crucial for cell division and cell growth (Seagull, 1989; Staiger and Lloyd, 1991). In order to detect gene interactions, some combinations between shape genes (especially *FASS*) and others have been tested, and most of the combinations suggest additive effects (Mayer, 1993; T. Berleth and R.A. Torres-Ruiz, unpublished data). In some cases *fs* seems to be epistatic to genes involved in the organization of the radial pattern (Scheres *et al.*, 1995). However, the respective radial mutants were represented by only one allele, which suggests that extreme caution is needed in assigning specific roles to these genes. Since the shape is altered while the pattern is preserved, one important lesson from the detailed analysis of *FASS* is that morphogenesis does not interfere with pattern formation (Torres-Ruiz and Jürgens, 1994). The manner whereby pattern formation and morphogenesis are integrated in order to generate the wild-type embryo remains unknown.

2.7 Strategies for cloning (apical) pattern genes

Several approaches have been adopted in order to clone pattern genes. Insertional mutagenesis is a technique whereby a known mobile or introduced DNA sequence becomes inserted into the gene of interest (Chapter 4, Lindsey *et al.*). This event not only leads to mutation of the gene but also allows 'direct' cloning by using the known inserted DNA as a probe to isolate the flanking DNA. In *Arabidopsis*, T-DNA tagging (Feldmann, 1991) and transposon tagging with heterologous systems from maize (e.g. Aarts *et al.*, 1993; Cardon *et al.*, 1993) have been used to exploit this strategy. The T-DNA tagging approach has led to the cloning of the first *Arabidopsis* embryo pattern gene *EMB30/GNOM* (Shevell *et al.*, 1994). Interestingly, this first example has also brought with it the first surprise. With regard to pattern genes in *Drosophila* which encode either transcription factors or elements of signal transduction pathways, it was believed that pattern genes in *Arabidopsis* would encode similar functions. However, *EMB30/GNOM* has been found to include a region with close similarities to SEC7 (called the 'Sec7 domain'), a protein of *Saccharomyces cerevisiae*. In *S. cerevisiae* this gene is involved in the transport of proteins through the Golgi apparatus. Confirming the results of genetic studies and cell-culture analysis, which suggest an *EMB30/GNOM* requirement during several stages of development (Baus *et al.*, 1986; Mayer *et al.*, 1993b; Meinke, 1985), its RNA is found throughout the life cycle of the plant (Shevell *et al.*, 1994). However, the functional significance of the Sec7 domain remains unknown in *Arabidopsis*. In yeast, deletion of the *SEC7* gene is lethal (Achstetter *et al.*, 1988), whereas in *Arabidopsis* mutations of this gene are not lethal at the cellular level since *emb30/gnom* mutants survive to the seedling stage (Mayer *et al.*, 1993b). This may indicate that, despite the sequence similarity, *EMB30/GNOM* does not share the same function in *Arabidopsis* as *SEC7* in yeast.

Unfortunately, the frequency of mutants obtained by T-DNA tagging is low (Feldmann, 1991). No tagged apical-pattern mutants resembling *gk*, *pep* and *lat* mutants have been reported. In these cases it has been necessary to follow other strategies. The extensive and dense molecular map of *Arabidopsis* allows the cloning of genes by chromosomal walking (also called positional or map-based cloning). This has led to the cloning of several genes (e.g. Giraudat *et al.*, 1992; Meyer *et al.*, 1994). In cases where a gene of interest lies in a region devoid of molecular markers, strategies have been developed to find new markers. Essentially, efforts are made to saturate the region of interest with molecular markers such that at least one marker is found which is so closely linked that chromosomal walking is not required. The term 'chromosomal landing' has recently been coined for this approach (Tanksley *et al.*, 1995). Two methods are currently used to find linked markers in plants with reasonable efficiency. The efficiency of the random amplified polymorphic DNA (RAPD) method (Williams *et al.*, 1990), as well as that of the amplified fragment length polymorphism method (Zabeau and Vos, 1993), have been demonstrated. Both methods include the use of the polymerase chain reaction (PCR) in conjunction with primers that amplify random loci.

In order to clone apical genes both chromosomal walking and chromosomal landing are currently being used. Mapping with morphological and molecular markers revealed that *GK* and *PEP* reside on different chromosomes. Markers on both sides of *GK* which were closely linked, could be isolated. Subsequently these were converted into cleaved amplified polymorphisms (CAPs) (Konieczny and Ausubel, 1993) in order to perform PCR fine mapping. Four segregating populations have been investigated (generated with different alleles of *GK*), which together include more than a thousand F_2 plants representing more than 2000 meiotic events. Two markers could be mapped; 0.3 cM on one side and 0.5 cM on the other side of *GK* (T. Fischer and R.A. Torres-Ruiz, unpublished data). The same markers have been used to isolate yeast artificial chromosomes (YACs) from *Arabidopsis* libraries which carry inserts as large as 400–500 kbp. Since 1 cM is equal to about 200 kbp in *Arabidopsis*, the isolated clones could well cover the *GK* region. Thus current research is following two approaches. Firstly, isolated fragments from these YACs are being used to test whether the YACs bridge the interval defined by both markers (and *GK*). Secondly, the same YACs are being used to isolate cDNAs which could again be used as polymorphic probes located nearer to *GK*, or which could partially represent *GK* itself. The subsequent identification of the gene will be facilitated with the aid of *GK* alleles used for genetic analysis (see above). In this context the isolated X-ray alleles could be of greater importance, since such mutants are known to harbour deletions or other chromosomal abnormalities.

Attempts to clone *PEP* have concentrated on isolating closely linked markers according to the chromosome landing strategy. This has been combined with bulk segregant analysis (Giovannoni *et al.*, 1991; Michelmore *et al.*, 1991) of populations segregating *pep* mutants. The test of a small batch of primers has already led to the isolation of at least two closely linked markers (G. Haberer and R.A. Torres-Ruiz, unpublished data). However, further work is needed in order to saturate the region where *PEP* maps. While the first steps towards cloning *PEP* will be very similar to those used for cloning *GK*, the identification of *PEP* will probably be different, since this gene is only represented by two alleles. The identification of *PEP* by sequence polymorphisms of the present alleles might be inadequate. Instead, in this case it would be necessary to identify the respective *PEP* clones by transformation to complement the mutant phenotype.

2.8 Further molecular studies for analysis of apical mutants

The closely linked markers and the mutant material allow molecular studies without the need to clone the genes. For instance, the markers can be used as tools for PCR-assisted double-mutant analysis of *GK* in combination with other pattern genes. This will be of major importance for those cases where the double mutant is not recognizable as the simple product of additivity. The mutant seedlings themselves suggest the possibility of differential display experiments using mini-preparations of mutant material.

2.8.1 *PCR-assisted double mutant analysis*

A few examples have shown that the identification of double mutants is not always easy. This is the case when testing *gk/pep* double mutants. Lines segregating both mutants are easily recognizable, since 7/16 of the progeny displays a mutant phenotype. In such lines no phenotypes other than those known for *gk* and *pep* could be observed (A. Lohner and R.A. Torres-Ruiz, unpublished data). In addition, *gk* lines show phenotypic variability (see above). One possible explanation for this variability is that *pep* has a similar, partially overlapping function to *gk*. Hence the question arises as to whether *gk/pep* double mutants always display the extreme phenotypes sometimes observed in *gk* single lines. The other possibility is that *pep* and *gk* are links in a biochemical chain, and double mutants vary to the same extent as *gk* single mutants. These possibilities cannot be discriminated by morphological inspection. On the other hand, it will take some time to isolate the respective genes which would allow direct analysis of (double) mutants by PCR or by *in situ* hybridization. Therefore, we have designed a PCR-assisted method which is based on closely linked markers which are polymorphic between two ecotypes of *Arabidopsis* (Figure 2.7). In addition to co-dominant linked markers the analysis requires the following 'components'. The mutants of the different genes must be generated in different ecotypes of *Arabidopsis* in order

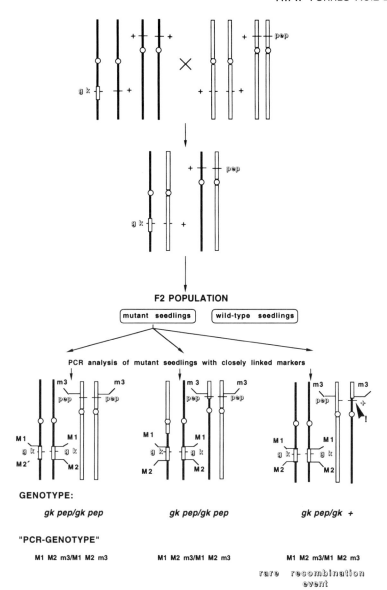

Figure 2.7. *PCR-assisted double-mutant analysis. Black line represents Niederzenz chromosome; open bar represents Landsberg chromosome; gk = gurke mutant allele; pep = pepino mutant allele; + represents corresponding wild-type alleles; M and m represent molecular markers from Niederzenz and Landsberg, respectively. Arrowhead points to a rare recombination event between closely linked molecular marker m3 and (mutant) pepino allele. For further explanation see text.*

to discriminate between markers linked to the wild-type gene of one ecotype and the mutant gene of the other ecotype. In the case of *gk* and *pep*, which have been induced in the same ecotype (Landsberg), it is necessary to collect appropiate recombinants from crosses of the ecotypes Landsberg (which carries the *gk* mutation) and Niederzenz (Figure 2.7). These were identified during the fine mapping process. The line with the *gk* mutation is a recombinant with Niederzenz markers linked to the mutant *gk* allele and Niederzenz markers linked to the wild-type *pep* allele. The line with the *pep* mutation is simply an isogenic Landsberg line with Landsberg markers linked to the mutant *pep* and the wild-type *gk* allele, respectively (Figure 2.7). These lines are crossed to produce F_1 plants. F_1 plants segregating double mutants in the F_2 generation will be recognized by a high frequency of mutant progeny. The mutants are inspected for their phenotype (e.g. rounded apical region vs. apical region with cotyledon rudiments) and subsequently analysed by PCR. With regard to the markers chosen for analysis, nearly all double recombinants will be reliably identified according to their PCR genotype, except for very rare new recombinants (Figure 2.7). By placing closely linked markers on each side of the appropriate gene, even rare recombination events which simulate double mutants will be detected.

This analysis is currently carried out for the *gk/pep* combination, but is also applicable to all other cases where the genes of interest are not cloned and identification of the double mutant is difficult.

2.8.2 Characterizing mutants by differential display analysis

The development of the apical region, especially the cotyledons, is likely to depend on an array of (interacting) genes. Genes that are required at very early stages of cotyledon development (such as *GK*, *PEP* and *LAT*) will probably precede the expression of others. The expression of genes at later stages might be directly or indirectly dependent on earlier acting genes. Therefore, in differential display experiments we have begun to identify such genes and to characterize further the apical defect mutants. In principle, such experiments could also lead to the isolation of defective genes in the respective mutants, although this is considered to be unlikely. For these experiments essentially the method of Sokolov and Prockop (1994) has been followed. Total RNA of wild-type and mutant seedlings was isolated after separation from DNA. mRNA was reverse-transcribed by using oligo-dT_{11} primers which were extended with adenine, cytosine or guanosine nucleosides. RAPD primers were then added to amplify random cDNA fragments which were subsequently separated on agarose gels. Bands which were visible in either wild-type, *gk* or *pep* preparations were isolated from the gels and again amplified with PCR. In Southern-hybridization experiments with labelled mRNA, some of the cDNA fragments gave a positive signal. A further control was prepared by amplifying cDNAs of known sequences using primers derived from the published sequence. A first set of clones has been sequenced and analysed for homologies in gene data collections. The result is summarized in Table 2.1. On the left-hand side of Table 2.1 the analysed clones are listed, and on the

Table 2.1. *Sequenced amplification products from differential display experiments*

Clone	Genomic DNA		Wild-type cDNA		*gurke* cDNA		*pepino* cDNA		Homologies
	1st	2nd	1st	2nd	1st	2nd	1st	2nd	
ddWTr1344	−	−	+	+	−	−	−	−	PS700 apoprotein A2 (AS, strong)
ddgkr1599/ r981.1	−	+ ?	+	+ ?	+	+	−	(+)	eIF4a/helicase (AS, weak)
ddWTYO13[a]	−	+	+	+	(+)	−	−	−	β-Conglycinin (soybean) (AS, very weak)
ddpepYO18	−	−	−	−	−	−	+	+ ?	EST(= expressed sequence tag)/*Brassica* (N; strong) identical to ddpepYO5c
ddpepYO5c	−	−	− ?	− ?	+	−	+	+ ?	EST/*brassica* (N; strong)
ddWTQO1	−	−	+	+	(+)	(+)	−	(+)	5.8S rRNA
ddgk/YO17b	−	−	− ?	+ ?	+	+ ?	+	+ ?	No homology found

[a]For detailed explanation see text.

right-hand side homologies (on DNA level = N and amino acid level = AS, respectively) are given. In the centre, + and − signs indicate whether amplification products of genomic DNA (control), wild-type cDNA, *gk* cDNA or *pep* cDNA were obtained. The results of two experiments are given. Obviously, additional clones must be sequenced in order to evaluate the usefulness of this approach. However, the initial results show that this approach may at least be useful for characterization of the mutants with respect to genes which are abundantly expressed. For instance, components of the photosynthetic apparatus or seed storage proteins appear to be under-represented in *pep* and *gk*, a result which correlates with their phenotypes. Although this approach is not devoid of false positives (5.8S rRNA), less abundant genes may also be found. The significance of these genes for the elaboration of the apical region of the plant seedling has yet to be determined.

2.9 Future and prospects

Genetic dissection of pattern formation and morphogenesis in the *Arabidopsis* embryo has revealed genes which are engaged from early on or at certain stages of development. Bearing in mind the essential interaction and coordination of the different components, an enormous amount of work remains to be done. The primary aims for approaching a mechanistic understanding of the fundamental events in embryology will be the identification and sequencing of these components, as well as the deciphering of their interactions.

Acknowledgements

We thank A. Lohner for technical assistance and A. Gierl for professional help with the scanning and processing of pictures. We also thank F. Assaad and S. Ploense for critically reading the manuscript. We gratefully acknowledge the support of this work by grants from the Deutsche Forschungsgemeinschaft.

References

Aarts, M.G., Dirkse, W.G., Stiekema, W.J. and Pereira, A. (1993) Transposon tagging of a male sterility gene in *Arabidopsis. Nature* **363**, 715–717.

Achstetter, T., Franzusoff, A., Field, C. and Schekman R. (1988) *SEC7* encodes an unusual, high molecular weight protein required for membrane traffic from the yeast Golgi apparatus. *J. Biol. Chem.* **263**, 11711–11717.

Barton, M.K. and Poethig, R.S. (1993) Formation of the shoot apical meristem in *Arabidopsis thaliana*: an analysis of development in the wild type and in the shoot meristemless mutant. *Development* **119**, 823–831.

Baus, A.D., Franzmann, L. and Meinke, D.W. (1986) Growth *in vitro* of arrested embryos from lethal mutants of *Arabidopsis thaliana. Theor. Appl. Genet.* **72**, 577–586.

Berleth, T. and Jürgens, G. (1993) The role of the *monopteros* gene in organising the basal body region of the *Arabidopsis* embryo. *Development* **118**, 575–587.

Cardon, G.H., Frey, M., Saedler, H. and Gierl, A. (1993) Mobility of the maize transposable element *En/Spm* in *Arabidopsis thaliana. Plant J.* **3**, 773–784.

Coen, E.S. and Meyerowitz, E. (1991) The war of the whorls: genetic interactions controlling flower development. *Nature* **353**, 31–37.

Crane, P.R., Friis, E.M. and Pedersen, K.R. (1995) The origin and early diversification of angiosperms. *Nature* **374**, 27–33.

van Engelen, F.A. and de Vries, S.C. (1992) Extracellular proteins in plant embryogenesis. *Trends Genet.* **8**, 66–70.

Feldmann, K.A. (1991) T-DNA insertion mutagenesis in *Arabidopsis:* mutational spectrum. *Plant J.* **1**, 71–82.

Giovannoni, J.J., Wing, R.A., Ganal, M.W. and Tanksley, S.D. (1991) Isolation of molecular markers from specific chromosomal intervals using DNA pools from existing mapping populations. *Nucleic Acids Res.* **23**, 6553–6558.

Giraudat, J., Hauge, B.M., Valon, C., Smalle, J., Parcy. F. and Goodman, H.M. (1992) Isolation of the *Arabidopsis ABI3* gene by positional cloning. *Plant Cell* **4**,1251–1261.

Goldberg, R.B., Barker, S.J. and Perez-Grau, L. (1989) Regulation of gene expression during plant embryogenesis. *Cell* **56**, 149–160.

Halperin, W. (1966) Alternative morphogenetic events in cell suspensions. *Am. J. Bot.* **53**, 443–453.

Ingham, P.W. (1988) The molecular genetics of embryonic pattern formation in *Drosophila. Nature* **335**, 25–34.

Johri, B.M. (1984) *Embryology of Angiosperms*. Springer Verlag, Berlin.

de Jong, A.J., Schmidt, E.D.L. and de Vries, S.C. (1993) Early events in higher-plant embryogenesis. *Plant Mol. Biol.* **22**, 367–377.

Jürgens, G. and Mayer, U. (1994) *Arabidopsis*. In: *Embryos. Colour Atlas of Development* (ed. J. Bard). Wolfe Publishing, London, pp. 7–21.

Jürgens, G., Mayer, U., Torres-Ruiz, R.A., Berleth, T. and Misera, S. (1991) Genetic analysis of pattern formation in the *Arabidopsis* embryo. *Development* **1** (Suppl.), 27–38.

Jürgens, G., Torres-Ruiz, R.A. and Berleth, T. (1994) Embryonic pattern formation in flowering plants. *Annu. Rev. Genet.* **28**, 351–371.

Kitano, H., Tamura, Y., Satoh, H. and Nagato, Y. (1993) Hierarchical regulation of organ differentiation during embryogenesis in rice. *Plant J.* **3**, 607–610.

Konieczny, A. and Ausubel, F.M. (1993) A procedure for mapping *Arabidopsis* mutations using co-dominant ecotype-specific PCR-based markers. *Plant J.* **4**, 403–410.

Liu, C.M., Xu, Z.H. and Chua, N.H. (1993) Auxin polar transport is essential for the establishment of bilateral symmetry during early plant embryogenesis. *Plant Cell* **5**, 621–630.

McWilliam, A.A., Smith, S.M. and Street. H.E. (1974) The origin and development of embryoids in suspension cultures of carrot (*Daucus carota*). *Ann. Bot.* **38**, 243–250.

Mansfield, S.G. and Briarty, L.G. (1991) Early embryogenesis in *Arabidopsis thaliana*. II. The developing embryo. *Can. J. Bot.* **69**, 461–476.

Mansfield, S.G. and Briarty, L.G. (1992) Cotyledon cell development in *Arabidopsis thaliana* during reserve deposition. *Can. J. Bot.* **70**, 151–164.

Mansfield, S.G., Briarty. L.G. and Erni, S. (1991) Early embryogenesis in *Arabidopsis thaliana*. I. The mature embryo sac. *Can. J. Bot.* **69**, 447–460.

Mayer, U. (1993) Entwicklungsgenetische Untersuchungen zur Musterbildung in Embryo der Blütenpflanze *Arabidposis thaliana*, Ph.D Thesis, Tübingen.

Mayer, U., Torres-Ruiz, R.A., Berleth, T., Miséra, S. and Jürgens, G. (1991). Mutations affecting body organization in the *Arabidopsis* embryo. *Nature* **353**, 402–407.

Mayer, U., Berleth, T., Torres-Ruiz, R.A., Miséra, S. and Jürgens, G. (1993a) Pattern formation during *Arabidopsis* embryo development. In: *Cellular Communication in Plants* (ed. R.M. Amasino). Plenum, New York, pp. 93–98.

Mayer, U., Büttner, G. and Jürgens, G. (1993b) Apical-basal pattern formation in the *Arabidopsis* embryo: studies on the role of the *gnom* gene. *Development* **117**, 149–162.

Meinke, D.W. (1985) Embryo-lethal mutants of *Arabidopsis thaliana:* analysis of mutants with a wide range of lethal phases. *Theor. Appl. Genet.* **69**, 543–552.

Meinke, D.W. and Sussex, I.M. (1979) Embryo-lethal mutants of *Arabidopsis thaliana*. A model system for genetic analysis of plant embryo development. *Dev. Biol.* **72**, 50–61.

Meyer, K., Leube, M.P. and Grill, E. (1994) A protein phosphatase 2C involved in ABA signal transduction in *Arabidopsis thaliana*. *Science* **264**, 1452–1455.

Meyerowitz, E.M. (1989) *Arabidopsis:* a useful weed. *Cell* **56**, 263–269.

Michelmore, R.W., Paran, I. and Kesseli, R.V. (1991) Identification of markers linked to disease-resistance genes by bulked segregant analysis: a rapid method to detect markers in specific genomic regions by using segregating populations. *Proc. Natl Acad. Sci. USA* **88**, 9828–9832.

Müller, A.J. (1963) Embryonentest zum Nachweis rezessiver Letalfaktoren bei *Arabidopsis thaliana*. *Biol. Zentralbl.* **82**, 133–163.

Nagato, Y., Kitano, H., Kamijima, O., Kikuchi, S. and Satoh, H. (1989) Developmental mutants showing abnormal organ differentiation in rice embryos. *Theor. Appl. Genet.* **78**, 11–15.

Neuffer, M.G. and Sheridan, W.F. (1980) Defective kernel mutants of maize. I. Genetic and lethality studies. *Genetics* **95**, 929–944.

Nüsslein-Volhard, C. (1991) Determination of the embryonic axes of *Drosophila*. *Development* **91** (Suppl.), 1–10.

Perez-Grau, L. and Goldberg, R.B. (1989) Soybean seed protein genes are regulated spatially during embryogenesis. *Plant Cell* **1**, 1095–1109.

Redei, G.P. (1975) *Arabidopsis* as a genetic tool. *Annu. Rev. Genet.* **9**, 111–127.

Rutishauser, A. (1969) *Embryologie und Fortpflanzungsbiologie der Angiospermen*. Springer Verlag, New York, Vienna.

Scheres, B., Wolkenfelt, H., Willemsen, V., Terlouw, M., Lawson, E., Dean. C. and Weisbeek, P. (1994) Embryonic origin of the *Arabidopsis* primary root meristem initials. *Development* **120**, 2475–2487.

Scheres, B., Di Laurenzio, L., Willemsen, V., Hauser, M.-T., Janmaat, K., Weisbeek, P. and Benfey, P. (1995) Mutations affecting the radial organisation of the *Arabidopsis* root display specific defects throughout the embryonic axis. *Development* **121**, 53–62.

Schiavone, F.M. and Racusen, R.H. (1990) Microsurgery reveals regional capabilities for pattern re-establishment in somatic carrot embryos. *Dev. Biol.* **141**, 211–219.

Schulz, R. and Jensen, W.A. (1968) *Capsella* embryogenesis: the egg, zygote, and young embryo. *Am. J. Bot.* **55**, 807–819.

Schwartz, B.W., Yeung, E.C. and Meinke, D.W. (1994) Disruption of morphogenesis and transformation of the suspensor in abnormal suspensor mutants of *Arabidopsis*. *Development* **120**, 3235–3245.

Schwarz-Sommer, Z., Huijser, P., Nacken, W., Saedler, H. and Sommer, H. (1990) Genetic control of flower development by homeotic genes in *Antirrhinum majus*. *Science* **250**, 931–936.

Seagull, R.W. (1989) The plant cytoskeleton. *CRC Crit. Rev. Plant Sci.* **8**, 131–167.

Sheridan, W.F. and Clark, J.K. (1993) Mutational analysis of morphogenesis of the maize embryo. *Plant J.* **3**, 347–358.

Shevell, D., Leu, W.-M., Gilmour, C.S., Xia, G., Felmann, K.A. and Chua, N.-H. (1994) *EMB30* is essential for normal cell division, cell expansion, and cell adhesion in *Arabidopsis* and encodes a protein that has similarity to *Sec7*. *Cell* **77**, 1051–1062.

Slack, J.M.W. (1993) *From Egg to Embryo*. Cambridge University Press, Cambridge.

Sokolov, B.P. and Prockop, D.J. (1994) A rapid and simple PCR-based method for isolation of cDNAs from differentially expressed genes. *Nucleic Acids Res.* **22**, 4009–4015.

Staiger, C.J. and Lloyd, C.W. (1991) The plant cytoskeleton. *Curr. Opin. Cell Biol.* **3**, 33–42.

Steeves, T.A. and Sussex, I.M. (1989) *Patterns in Plant Development*. Cambridge University Press, Cambridge.

Sterk, P., Booij, H., Schellekens, G.A., Van Kammen, A. and de Vries, S.C. (1991) Cell-specific expression of the carrot EP2 lipid transfer protein gene. *Plant Cell* **3**, 907–921.

Takhtajan, A.L. (1980) Outline of the classification of flowering plants (Magnoliophyta). *Bot. Rev.* **46**, 225–259.

Tanksley, S.D., Ganal, M.W. and Martin, G.B. (1995) Chromosome landing: a paradigm for map-based gene cloning in plants with large genomes. *Trends Genet.* **11**, 63–68.

Torres-Ruiz, R.A. and Jürgens, G. (1994) Mutations in the *FASS* gene uncouple pattern formation and morphogenesis in *Arabidopsis* development. *Development* **120**, 2967–2978.

Tykarska, T. (1976) Rape embryogenesis. I. The proembryo development. *Acta Soc. Bot. Pol.* **45**, 3–15.

Tykarska, T. (1979) Rape embryogenesis. II. Development of the embryo proper. *Acta Soc. Bot. Pol.* **48**, 391–421.

Weigel, D., Alvarez, J., Smyth, D.R., Yanofsky, M.F. and Meyerowitz, E.M. (1992) *LEAFY* controls floral meristem identity in *Arabidopsis*. *Cell* **69**, 841–859.

Williams, J.G.K., Kubelik, A.R., Livak, K.J., Rafalski, J.A. and Tingey, S.V. (1990) DNA polymorphisms amplified by arbitrary primers are useful as genetic markers. *Nucleic Acids Res.* **18**, 6531–6535.

Zabeau, M. and Vos, P. (1993) *Eur. Pat. Appl. No. 92402629.7.*

3

Embryo-defective mutants of *Arabidopsis*: cellular functions of disrupted genes and developmental significance of mutant phenotypes

D.W. Meinke

3.1 Introduction

The study of plant development has changed dramatically in recent years, from an initial emphasis on descriptive and experimental studies involving many different species, to detailed genetic dissections of selected model organisms such as maize (Neuffer *et al.*, 1995) and *Arabidopsis* (Meyerowitz and Somerville, 1994). A common strategy in recent studies has been to isolate and characterize developmental mutants, determine the normal functions of disrupted genes, and then attempt to understand the complex molecular interactions that occur between different gene products. Embryogenesis has become a popular subject for genetic analysis because it plays such a critical role in the life cycle of higher plants (Meinke, 1995). Large numbers of genes must be expressed as the zygote develops into a multicellular embryo capable of surviving desiccation and germinating to produce a viable seedling. One long-term goal of research in plant developmental genetics has therefore been to isolate and characterize a wide range of genes with essential functions during embryogenesis.

Genetic approaches to the study of plant embryo development have recently been reviewed in detail (Goldberg *et al.*, 1994; Jürgens, 1994; Jürgens *et al.*, 1994; Meinke, 1991a, b, 1994, 1995; West and Harada, 1993). Several other chapters of this book also deal with genetic analysis of plant embryo development. The purpose of this article is not to duplicate those reviews, but rather to present an update on recent studies involving embryo-defective (*emb*) mutants of *Arabidopsis*. Several *EMB* genes disrupted by transposon tagging and T-DNA insertional mutagenesis have now been cloned and assigned a cellular function. The picture emerging from analysis of these cloned genes and the entire collection of embryo-defective mutants is that:

(a) negative regulation of developmental potential is an important feature of plant embryogenesis;

(b) embryonic pattern mutants are not necessarily defective in genes that function to regulate pattern formation directly;

(c) many essential cellular functions are genetically redundant and therefore protected to a great extent from the effects of deleterious mutations; and

(d) stored maternal gene products may play an important role in supporting the growth of embryos at early stages of development.

In addition, molecular studies of embryo-defective mutants are beginning to provide insights into the genetic control of basic cellular functions in higher plants.

3.2 Diversity of available mutants

Over 500 embryo-defective mutants of *Arabidopsis* are currently being analysed in a number of laboratories world-wide. These mutants were isolated following either chemical mutagenesis, X-irradiation, T-DNA insertional mutagenesis, or gene tagging with a maize transposable element. Large collections of mutants are being maintained in the laboratory at Oklahoma State University, and in the laboratories of Gerd Jürgens and colleagues at the University of Tubingen, Robert Goldberg and colleagues at the University of California, and Michel Delseny at the University of Perpignan. At least 15 other laboratories are focusing on specific mutants with interesting defects in embryogenesis. The high frequency of *emb* mutants identified following seed mutagenesis is consistent with the presence of large numbers of genes with essential functions at this stage of the life cycle. Many of these mutants were identified by screening immature siliques for the presence of 25% defective seeds following self-pollination. Others were identified at the seedling stage by screening for defects indicative of a disruption of normal embryogenesis.

My research group initially referred to these mutants as 'embryonic lethals', following the tradition of Müller (1963), because they often exhibited dramatic abnormalities at early stages of embryogenesis. This was an unfortunate decision because some mutant embryos were later found to continue development to a limited extent, and even to germinate at maturity. More recently, the nomenclature has been changed and the term 'embryonic defective' used to describe the entire collection of mutants defective in embryogenesis. Included among the 250 mutants maintained in our collection are fascinating examples of defects in growth, morphogenesis, cell division, cell differentiation, and pattern formation. Some of these mutants are likely to represent strong alleles of mutants known in other laboratories for their aberrations in vegetative development. Others are identical to pattern mutants isolated elsewhere at the seedling stage. Some mutants may even be weak alleles of female gametophytic factors. The common feature that links all of these mutants is that embryo development becomes abnormal at some point after fertilization and before the onset of germination. Examples of mutant phenotypes are shown in Figure 3.1.

3.3 Number of target genes

How many genes in *Arabidopsis* can readily mutate to give an embryo-defective phenotype? My laboratory chose to address this question first by comparing the frequency of embryo-defective mutants with that of other mutants identified

Figure 3.1. *Examples of embryo-defective mutants of* Arabidopsis. *(a) Portion of a heterozygous silique containing immature mutant (white) and normal (green) seeds. (b) Mixture of* fusca *(dark brown) and normal (light brown) seeds at maturity. (c) Immature* fusca *embryos showing inappropriate accumulation of anthocyanin in cotyledons. (d) Cotyledons of a wild-type seedling germinated in culture. (e) Cotyledons of a* leafy cotyledon *seedling produced by precocious germination of an immature mutant embryo in culture.*

following *Agrobacterium*-mediated seed transformation, and then by mapping the chromosomal locations of mutant genes and determining the frequency of duplicate alleles. The results obtained by both methods were consistent with an estimate that approximately 500 target genes may produce an embryo-defective phenotype in *Arabidopsis* (Franzmann *et al.*, 1995). To date we have obtained recombination data with linked visible markers for 169 mutants defective in embryogenesis, identified several putative chromosomal translocations among T-DNA insertional mutants, and placed 110 *emb* loci on the genetic map of *Arabidopsis* (Castle *et al.*, 1993; Franzmann *et al.*, 1995; Patton *et al.*, 1991). Eleven genes are represented by two different alleles each and three genes are represented by three different alleles each. Although we have not begun to approach saturation for embryo-defective mutants, it is estimated that one out of every five new mutants identified at this time should be allelic to a mutant already on the map. We have therefore established a solid foundation for future studies on the identification of genes with essential functions during plant embryogenesis.

3.4 Cellular functions of disrupted genes

Many embryo-defective mutants are likely to be altered in basic cellular functions required for normal growth and development. Distinguishing these 'housekeeping' mutants from those with defects in critical regulatory functions represents a major challenge for investigators in this field. My research group has chosen to examine a representative sample of mutants with different patterns of abnormal development, attempt to uncover the molecular basis of abnormal development in these mutants, and then draw some general conclusions about the relationship between gene function and normal development. This approach has finally proved to be successful. Several fascinating examples of mutants defective in basic cellular and developmental functions have now been identified. The history and developmental significance of these discoveries are considered in the remainder of this chapter.

3.4.1 *Basic cellular metabolism*

Defects in basic cellular metabolism should result in lethality if duplicate genes or biochemical pathways are unable to rescue the developing embryo. The first example of an embryo-defective mutant with a known metabolic defect was the *bio1* auxotroph of *Arabidopsis* (mutant 122G-E), isolated and characterized in my laboratory (Baus *et al.*, 1986). Mutant *bio1* embryos become abnormal at the heart stage of development but develop into normal plants when grown in the presence of supplemental biotin (Schneider *et al.*, 1989). Mutant embryos can also be rescued by desthiobiotin and 7,8-diaminopelargonic acid, two known intermediates of biotin synthesis in bacteria, but not by their immediate precursor, 7-keto-8-aminopelargonic acid (Shellhammer, 1991; Shellhammer and Meinke, 1990). The relationship between these biosynthetic intermediates is summarized

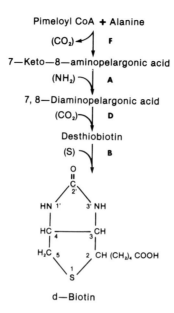

Figure 3.2. *Overview of biotin synthesis in* E. coli. *Letters F, A, D and B represent genes required for different steps in the pathway.*

in Figure 3.2. Mutant seeds contain normal levels of biotin in the heterozygous seed coat but reduced levels of biotin in the embryo (Shellhammer and Meinke, 1990). On the basis of these initial observations, we proposed that the biotin synthetic pathway was conserved between plants and bacteria, and that *bio1* embryos were defective in the synthesis of 7,8-diaminopelargonic acid from 7-keto-8-aminopelargonic acid, a step catalysed in bacteria by the *bioA* gene. This model has recently been confirmed by transforming *bio1* plants with a normal copy of the *bioA* gene from *Escherichia coli* and noting a dramatic reduction in the symptoms of biotin deficiency (David Patton and Eric Ward, personal communication). This experiment provides not only a direct confirmation of the biochemical defect in *bio1* embryos, but also a novel example of a plant mutant being rescued through the introduction of a cloned bacterial gene. Whether other auxotrophs are included in our collection of mutants remains to be determined.

3.4.2 *Cell structure and organization*

Another embryo-defective mutant, known as *emb30* (Meinke, 1994) or *gnom* (Mayer *et al.*, 1993), exhibits a different type of defect in a basic cellular function. This mutant has an interesting history that illustrates how the same locus can be identified through screens performed at both the embryo and seedling stages. The first allele (112A-2A) was identified in my laboratory based on the altered shape of mutant embryos at maturity (Meinke, 1985). The main focus of initial studies was on the fused appearance of embryonic cotyledons and the rootless phenotype of mutant plants rescued in culture (Baus *et al.*, 1986; Patton and

Meinke, 1990). Many additional alleles were subsequently isolated and character-ized in detail by Gerd Jürgens and colleagues (Mayer *et al.*, 1991). The mutant phenotype was then interpreted as a defect in embryonic pattern formation, specifically a deletion of the apical and basal portions of the developing plant. Mutant embryos were also found to exhibit altered patterns of cell division early in development (Mayer *et al.*, 1993). This led to a model in which *EMB30/GNOM* was viewed as playing an important role in directing the initial asymmetrical division of the zygote during early embryogenesis (Jürgens, 1994; Jürgens *et al.*, 1994).

The recent cloning and molecular characterization of *EMB30/GNOM* (Shevell *et al.*, 1994) does not readily support such a direct role for this gene in regulating embryonic pattern formation. Instead, *EMB30* appears to perform a more general cellular function throughout the life cycle. The defect in cell division and pattern formation observed in mutant embryos may simply be one manifestation of a more general alteration of cell division patterns throughout development. Sequence analysis of *EMB30* has revealed several regions of homology to the *SEC7* gene of yeast, which facilitates intracellular protein transport through the Golgi. The original *emb30-1* allele has a single base-pair substitution within a small domain that is conserved among putative *SEC7* homologues from other organisms. This has led to a revised model in which *EMB30/GNOM* plays a role in processing or transporting materials to the cell surface. Loss of normal gene function results in a variety of defects, some of which resemble changes in pattern formation. It therefore appears that mutant embryos are not disrupted in a gene that functions directly or solely to regulate pattern formation. Whether such genes exist in higher plants, and whether they can be readily identified by loss-of-function mutations, are questions that have yet to be answered.

3.4.3 *Signal transduction pathways*

The *fusca* class of embryo-defective mutants provides an interesting example of an alteration in a complex signal transduction pathway. These mutants were first identified by Müller (1963) on the basis of inappropriate accumulation of anthocyanins in developing cotyledons. Mutant embryos typically germinate to produce defective seedlings that fail to complete the normal life cycle. Anthocyanin accumulation is an indirect result of the mutation, and not the primary cause of seedling lethality. This conclusion is based on the observation that *fusca/transparent testa* double-mutant embryos still exhibit the same defects in seedling development even though they lack anthocyanins (Castle and Meinke, 1994). Twelve *fusca* loci have been identified through the combined efforts of several research groups (Castle and Meinke, 1994; Miséra *et al.*, 1994). Phenotypes of mutant embryos and seedlings differ according to locus and mutant allele. Strong alleles are generally lethal at the seedling stage. The *fusca* seed phenotype is illustrated in Figure 3.1.

My research group became interested in the *fusca* phenotype when several mutants of this type were found among populations of transgenic plants produced following *Agrobacterium*-mediated seed transformation (Castle and Meinke, 1994; Errampalli *et al.*, 1991; Feldmann, 1991). These mutants became even more interesting when it was discovered that some of the *de-etiolated* (*det*) and *constitutive photomorphogenic* (*cop*) mutants, known previously for their peculiar defects in light responses, were actually *fusca* mutants (Castle and Meinke, 1994). Several of these genes have now been cloned and sequenced, namely *FUS1/COP1* (Deng *et al.*, 1992), *FUS2/DET1* (Pepper *et al.*, 1994), *FUS6/COP11* (Castle and Meinke, 1994) and *FUS7/COP9* (Wei *et al.*, 1994). *DET1* and *COP9* appear to encode novel proteins not represented in existing databases. The *COP1* gene product has an N-terminal zinc-binding domain, an internal coiled-coil helix structure, and a C-terminal domain with homology to the B-subunit of hetero-trimeric G-proteins (Deng *et al.*, 1992). These regions of homology are consistent with a model in which the *COP1* protein interacts directly with both regulatory proteins and target DNA sequences. *FUS6* initially appeared to encode a novel protein because the only related sequence found in the database was a random partial cDNA from rice (Castle and Meinke, 1994). This view has now changed with the discovery that *FUS6* is closely related to a human sequence (*GPS1*) that was isolated on the basis of its ability to rescue yeast mutants disrupted in G-protein-mediated signal transduction pathways (L. Castle, personal communi-cation). This observation is consistent with the emerging view that G-proteins play an important role in light signal transduction pathways in plants (Bowler and Chua, 1994). *FUS6* may therefore act downstream of cGMP in a signal transduction network that interprets a variety of environmental, hormonal and nutritional signals during embryonic maturation and early germination. Regardless of the specific role played by this gene in plant development, it should be apparent that the analysis of a small group of embryo-defective mutants has already provided valuable insights into complex mechanisms of signal transduction pathways in higher plants.

3.4.4 *Replication, transcription and translation*

Several embryo-defective mutants of *Arabidopsis* have recently been found with lesions in cellular factors required for replication, transcription and translation. The *prolifera* mutant identified in a screen for transposon-induced mutations appears to be a leaky megagametophytic factor that causes both female sterility and occasional embryonic lethality (P. Springer and R. Martienssen, personal communication). Interestingly, the *PRL* gene shows homology with the yeast *MCM3* family of genes involved in the initiation of DNA replication. Normal expression of this gene is likely to be required for the completion of megagametogenesis. The mutant allele may have sufficient residual activity to allow a limited amount of embryo development in some mutant seeds. The discovery that the transposon responsible for the *prolifera* mutation inserted within an intron, and therefore was unlikely to result in a null phenotype, is consistent with this model. Another *EMB* gene tagged with a maize transposable element has recently been shown to encode a ribosomal protein (R. Tsugeki and

N. Fedoroff, personal communication). The squence identified is most closely related to a ribosomal protein (RPS16) in *Neurospora* that is encoded by a nuclear gene but localized in the mitochondrion. Lethality may therefore result from a failure of mutant embryos to produce within the mitochondrion the materials required for the completion of normal growth and development.

Another mutant examined in our laboratory appears to be defective in an essential spliceosome assembly factor. This mutant was initially chosen for detailed analysis because it provided a particularly striking example of abnormal growth of the suspensor, a filamentous structure that supports growth of the early embryo proper but then undergoes programmed cell death and is not present at maturity (Yeung and Meinke, 1993). In many embryo-defective mutants, developmental abnormalities in the embryo proper are accompanied by renewed growth and cellular transformation of the suspensor. Examples of this mutant phenotype are shown in Figure 3.3. Mutants with particularly large suspensors are known as *sus* mutants. Two such loci (*sus1* and *sus2*), defined by three alleles each, have been examined in most detail (Schwartz *et al.*, 1994). Another mutant with a similar phenotype has also been described elsewhere (Yadegari *et al.*, 1994). Plasmid rescue has been used in my laboratory to recover plant sequences flanking the T-DNA insert in a tagged *sus2* allele known originally as *emb177* (Errampalli *et al.*, 1991). Selected regions of the corresponding wild-type gene have been sequenced (B. Schwartz, unpublished data) and shown to be closely related to a yeast gene (*PRP8*) required for spliceosome assembly (Brown and Beggs, 1992). Although yeast mutants defective in this gene have been examined in detail, and mutants disrupted in other splicing factors have been identified in other organisms,

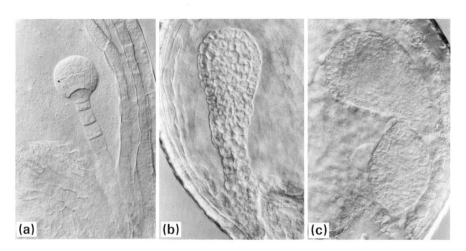

(a)   (b)   (c)

Figure 3.3. *Abnormal suspensor mutants of* Arabidopsis. *Phenotypes of immature seeds examined with Nomarski optics. (a) Wild-type suspensor and embryo proper at a globular stage of development. (b) Mutant embryo (sus1) with columnar suspensor. (c) Mutant embryo (sus3) with massive suspensor.*

mutants altered in *PRP8* have not previously been found in a multicellular eukaryote. Molecular characterization of *SUS2* should therefore provide a unique opportunity to dissect RNA processing in plants from a genetic perspective.

3.4.5 *Regulation of development*

The original goal in working with embryo-defective mutants of *Arabidopsis* was to identify genes that played a direct role in the regulation of plant embryo development (Meinke and Sussex, 1979). This goal has finally been realized through the isolation and characterization of two classes of embryo-defective mutants with particularly unusual phenotypes. The first class is known as *leafy cotyledon* mutants, because developing cotyledons of mutant embryos fail to activate a wide range of embryo-specific pathways and instead acquire features that are normally restricted to vegetative leaves. At least three different genes can mutate to give this dramatic phenotype. The *lec1-1* allele isolated following *Agrobacterium*-mediated seed transformation was the first mutant described (Meinke, 1992). The phenotype of this mutant is shown in Figure 3.1. A second allele with an identical phenotype (*lec1-2*) was subsequently identified (Meinke *et al.*, 1994; West *et al.*, 1994). Two other mutants (*lec2* and *fus3*) with related but distinct phenotypes were also found (Baumlein *et al.*, 1994; Keith *et al.*, 1994; Meinke *et al.*, 1994). The mutant phenotype has been interpreted as either a partial homeotic transformation of cotyledons to leaf-like structures (Meinke, 1992) or a heterochronic shift in development of the mutant embryo (Keith *et al.*, 1994). My research group prefers the homeotic interpretation because it seems most consistent with the evolution of cotyledons as modified leaves. According to this model, *LEC* genes positively regulate many features of cell differentiation during embryonic maturation, but also negatively regulate the developmental potential of embryonic cotyledons (Meinke *et al.*, 1994). Expression of the wild-type gene is required to complete embryo morphogenesis, suppress leafy traits in cotyledons, suppress precocious germination, and activate late embryogenesis programmes such as storage product accumulation and the acquisition of desiccation tolerance. In the absence of normal gene function, mutant cotyledons revert to a basal developmental state that is somewhat leaf-like. A similar conclusion has been drawn with regard to the basal developmental state of floral organs, which produce leafy structures in the absence of regulatory factors required for organ specification (Coen and Meyerowitz, 1991). The normal sensitivity of *lec* embryos to exogenous ABA and the phenotypes of *lec abi3* double mutants indicate that *LEC* and *ABI3* probably function in separate pathways (Meinke *et al.*, 1994). *LEC1* therefore appears to perform a major and previously unknown role in regulating morphogenesis and cell differentiation during plant embryo development.

The *twin* mutants provide a second example of a dramatic phenotype that may result from the loss of a critical regulatory function during embryogenesis. The first mutant of this type (*twn1*) was identified in the laboratory as part of a larger project designed to characterize late embryo-defective mutants of *Arabidopsis* (Vernon and Meinke, 1994, 1995). As indicated by the name of this locus, homozygous mutant seeds often contain twin embryos that germinate to produce twin seedlings. Some mutant embryos and seedlings exhibit a variety of defects in morphogenesis, while others are surprisingly normal. Genetic and developmental studies have shown that *twin* is a pleiotropic mutant, with incomplete penetrance, that is disrupted in a single mapped gene (Vernon and Meinke, 1994). The most intriguing and informative aspect of the mutant phenotype concerns the origin of the secondary embryo. In contrast to most examples of spontaneous polyembryony in higher plants, twin embryos in this mutant develop from the suspensor of the original embryo. One such twin embryo is shown in Figure 3.4. The secondary embryo appears in some cases to recapitulate the entire spectrum of cellular and developmental events that are normally restricted to the embryo proper. This observation is consistent with our model that the suspensor in *Arabidopsis* and other angiosperms has a

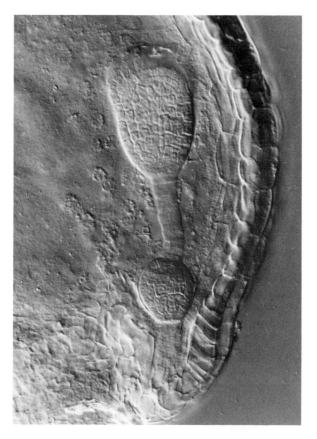

Figure 3.4. *Twin mutant of* Arabidopsis. *Nomarski optics of immature mutant seed containing heart-stage embryo proper and globular-stage secondary embryo produced from the suspensor.*

developmental potential that far exceeds its normal developmental fate, and that this developmental potential is ordinarily inhibited by the embryo proper (Marsden and Meinke, 1985; Yeung and Meinke, 1993). The *TWN* gene appears to play a central role in maintaining suspensor cell identity, suppressing embryogenic potential of the suspensor, and regulating morphogenesis of the embryo proper (Vernon and Meinke, 1994). Additional mutants with similar phenotypes have recently been found in the laboratories of Chris Somerville (J. Zhang, personal communication) and Gerd Jürgens (S. Ploense, personal communication). On the basis of preliminary map data and allelism tests, these mutants appear to have identified at least one additional gene with a related function during embryo development. It therefore appears that, with both the *leafy cotyledon* and *twin* mutants, the initial discovery of an intriguing phenotype has allowed the identification of small networks of genes that may function in a co-ordinated manner to regulate important developmental events during plant embryogenesis.

3.5 Developmental significance of mutant phenotypes

The representative studies outlined above illustrate how the analysis of embryo-defective mutants can result in the discovery of interesting genes with important functions during plant growth and development. Some of these genes perform essential housekeeping functions throughout the life cycle, while others appear to play a more direct role in the regulation of embryo development. Many additional *EMB* genes are likely to be identified and cloned in the future, allowing even more detailed genetic analysis of basic cellular functions in plants. What specifically have we learned about plant development as a result of studying these large collections of mutants? I believe that we have gained valuable insights into at least one unexpected feature of late embryogenesis, two important features of early embryogenesis, and several general questions concerning the relationship between gene function and plant development.

3.5.1 *Negative regulation of developmental potential*

One important discovery attributed to the analysis of embryo-defective mutants, is that cotyledons have the potential not only to germinate precociously, which had been realized for many years, but also to become transformed into leaf-like structures. The activation of embryo-specific maturation programmes at the heart stage of development may therefore be coupled with a repression of the underlying potential of embryonic cotyledons to become leaf-like. A similar repression of developmental potential appears to occur early in development between the embryo proper and suspensor. Whether this repression is coupled with activation of developmental programmes in the embryo proper has yet to be determined. Negative regulation of plant development has also emerged as a theme from genetic, molecular and biochemical studies of photomorphogenesis (Bowler and Chua, 1994). Morphogenesis and cell differentiation in plants may therefore require both limited activation of tissue-specific programmes and general repression of underlying developmental potential. Although the totipotency of plant cells has long been recognized, the role of specific regulatory genes in repressing this developmental potential during embryogenesis has largely been overlooked.

3.5.2 *Importance of stored maternal gene products*

Another conclusion that has begun to emerge from the analysis of embryo-defective mutants is that stored maternal gene products may play an important role in supporting the initial growth and development of a plant embryo. We had previously assumed that stored gene products played a relatively minor role in early embryogenesis, based primarily on the apparent scarcity of maternal-effect embryonic lethals in *Arabidopsis*. The phenotype of developing embryos in such mutants should be determined by the genotype of diploid maternal cells in direct contact with the egg and developing megagametophyte. Thus, heterozygotes should produce 100% normal seeds, whereas homozygotes should produce 100% defective seeds. Mutants of this type have not yet been described in *Arabidopsis*, despite extensive screening of transgenic populations. This observation is difficult to reconcile with the apparent scarcity of embryo-defective mutants in which development becomes arrested immediately after fertilization. It would appear that mutants of this type should be relatively common if an essential cellular function has been disrupted.

Phenotypes of mutant embryos are difficult to interpret without knowing whether the mutant allele being studied is a null, whether the gene product modifies a diffusible substance that can be provided by surrounding maternal tissues, or whether the gene in question is part of a multigene family in which related members are expressed at different stages of development. In the case of *bio1*, there may be only a single copy of this gene, the mutant may in fact be a null, and the biotin requirements of young embryos may be met by a heterozygous seed coat rich in biotin. Mutant embryos may then become arrested later in development when biotin becomes a limiting factor. Many other mutants, however, lack essential factors that cannot readily be provided by the seed coat.

Why then do many tagged mutant embryos reach a globular stage of development when there is probably little residual gene activity to support growth of the early embryo? Perhaps the answer is that development of the proembryo is supported not only by gene expression after fertilization, but also by maternal gene products deposited in the egg. Maternal-effect lethals would then be rare because the missing gene product could be supplied by the developing embryo, but zygotic lethals would also be rare because many essential functions could be supported for at least a short time by low levels of stored gene products. This duplication of function might also protect the young embryo from the effects of deleterious mutations. If this model is eventually found to be correct, it would provide an interesting contrast with embryogenesis in *Drosophila*, where there is very little evidence of zygotic gene expression prior to the blastoderm stage of development, and where maternal gene products play an important role in both cellular functions and morphogenesis. Whether plants have a small number of stored gene products that perform regulatory functions that are not duplicated by the embryo is a potentially significant question that has yet to be answered.

3.5.3 *Genomic redundancy of essential functions*

There are likely to be approximately 10 000–15 000 functional genes in *Arabidopsis*. This estimate is based on the known genome size, the average size of a typical gene, and the estimated spacing between genes. At least 20–30% of these genes are likely to be expressed during embryo development. This produces a conservative estimate that several thousand genes are expressed during embryo development. All of these genes should in theory be targets for mutagenesis. The observation that only about 500 genes appear to mutate readily to give an embryo-defective phenotype in *Arabidopsis* suggests that many genes expressed during embryogenesis cannot be identified by loss-of-function mutations. The most logical explanation is that many essential functions in *Arabidopsis* are encoded by duplicated genes. A similar conclusion has recently been drawn from large-scale genome-sequencing efforts in model organisms such as *Saccharomyces* and *Caenorhabditis*. There may also be considerable redundancy of metabolic pathways in plants. As a result, many essential cellular and developmental processes in plants may be difficult to analyse from a genetic perspective. Therefore, one challenge for the future will be to use information obtained from large-scale sequencing efforts to identify genes with important developmental functions that have escaped detection in screens for loss-of-function mutations. From an evolutionary perspective, one potential advantage of this genomic and biochemical duplication of essential cellular functions may be that developing embryos are less sensitive to the effects of deleterious mutations. In any case, it appears that even plants with small genomes contain a surprisingly high level of genetic redundancy. The challenge for the future will be to identify not only those unique genes whose loss of function results in a mutant phenotype, but also the large numbers of duplicated genes that contribute to essential cellular and developmental programmes throughout embryo development.

Acknowledgements

I thank the members of my research group who contributed to this research, including David Patton, Linda Franzmann, Elizabeth Yoon, Aynsley Kealiher (mapping project), Joe Shellhammer (biotin auxotroph), Deena Errampalli (transgenic analysis), Linda Castle (*fusca* mutants), Brian Schwartz (*suspensor* mutants), Dan Vernon (*twin* mutant), Todd Nickle and the many undergraduates who assisted with screening and maintenance of plants. Ken Feldmann (University of Arizona) and Ed Yeung (University of Calgary) made many contributions to this project. I also thank those colleagues who shared unpublished data for this article. Research in my laboratory has been supported by grants from the

National Science Foundation (Developmental Biology, Special Projects, and EPSCoR Programs), the US Department of Agriculture (Competitive Grants Program), and the S.R. Noble Foundation (Plant Biology Division) in Ardmore, Oklahoma.

References

Baumlein, H., Miséra, S., Luersen, H., Kolle, K., Horstmann, C., Wobus, U. and Müller, A.J. (1994) The *FUS3* gene of *Arabidopsis thaliana* is a regulator of gene expression during late embryogenesis. *Plant J.* **6**, 379–387.

Baus, A.D., Franzmann, L. and Meinke, D.W. (1986) Growth *in vitro* of arrested embryos from lethal mutants of *Arabidopsis thaliana*. *Theor. Appl. Genet.* **72**, 577–586.

Bowler, C. and Chua, N.-H. (1994) Emerging themes of plant signal transduction. *Plant Cell*, **6**, 1529–1541.

Brown, J.D. and Beggs, J.D. (1992) Roles of PRP8 protein in the assembly of splicing complexes. *EMBO J.* **11**, 3721–3729.

Castle, L.A. and Meinke, D.W. (1994) A *FUSCA* gene of *Arabidopsis* encodes a novel protein essential for plant development. *Plant Cell* **6**, 25–41.

Castle, L.A., Errampalli, D., Atherton, T.L., Franzmann, L.H., Yoon, E.S. and Meinke, D.W. (1993) Genetic and molecular characterization of embryonic mutants identified following seed transformation in *Arabidopsis*. *Mol. Gen. Genet.* **241**, 504–514.

Coen, E.S. and Meyerowitz, E.M. (1991) The war of the whorls: genetic interactions controlling flower development. *Nature* **353**, 31–37.

Deng, X.-W., Matsui, M., Wei, N., Wagner, D., Chu, A.M., Feldmann, K.A. and Quail, P.H. (1992) *COP1*, an *Arabidopsis* regulatory gene, encodes a protein with both a zinc-binding motif and a GB homologous domain. *Cell* **71**, 791–801.

Errampalli, D., Patton, D., Castle, L., Mickelson, L., Hansen, K., Schnall, J., Feldmann, K. and Meinke, D. (1991) Embryonic lethals and T-DNA insertional mutagenesis in *Arabidopsis*. *Plant Cell* **3**, 149–157.

Feldmann, K.A. (1991) T-DNA insertion mutagenesis in *Arabidopsis*: mutational spectrum. *Plant J.* **1**, 71–82.

Franzmann, L.H., Yoon, E.S. and Meinke, D.W. (1995) Saturating the genetic map of *Arabidopsis thaliana* with embryonic mutations. *Plant J.* **7**, 341–350.

Goldberg, R.B., DePaiva, G. and Yadegari, R. (1994) Embryogenesis: zygote to seed. *Science* **266**, 605–614.

Jürgens, G. (1994) Pattern formation in the embryo. In: Arabidopsis (eds E.M. Meyerowitz and C.R. Somerville). Cold Spring Harbor Laboratory Press, Cold Spring Harbor, pp. 297–312.

Jürgens, G., Torres Ruiz, R.A. and Berleth, T. (1994) Embryonic pattern formation in flowering plants. *Annu. Rev. Genet.* **28**, 351–371.

Keith, K., Kraml, M., Dengler, N.G. and McCourt, P. (1994) *fusca3*: a heterochronic mutation affecting late embryo development in *Arabidopsis*. *Plant Cell* **6**, 589–600.

Marsden, M.P.F. and Meinke, D.W. (1985) Abnormal development of the suspensor in an embryo-lethal mutant of *Arabidopsis thaliana*. *Am. J. Bot.* **72**, 1801–1812.

Mayer, U., Torres-Ruiz, R.A., Berleth, T., Miséra, S. and Jürgens, G. (1991) Mutations affecting body organization in the *Arabidopsis* embryo. *Nature* **353**, 402–407.

Mayer, U., Buttner, G. and Jürgens, G. (1993) Apical-basal pattern formation in the *Arabidopsis* embryo: studies on the role of the *gnom* gene. *Development* **117**, 149–162.

Meinke, D.W. (1985) Embryo-lethal mutants of *Arabidopsis thaliana*: analysis of mutants with a wide range of lethal phases. *Theor. Appl. Genet.* **69**, 543–552.

Meinke, D.W. (1991a) Perspectives on genetic analysis of plant embryogenesis. *Plant Cell* **3**, 857–866.

Meinke, D.W. (1991b) Embryonic mutants of *Arabidopsis thaliana. Dev. Genet.* **12**, 382–392.

Meinke, D.W. (1992) A homoetic mutant of *Arabidopsis thaliana* with leafy cotyledons. *Science* **258**, 1647–1650.

Meinke, D.W. (1994) Seed development in *Arabidopsis thaliana*. In: Arabidopsis (eds E.M. Meyerowitz and C.R. Somerville). Cold Spring Harbor Laboratory Press, Cold Spring Harbor, pp. 253–295.

Meinke, D.W. (1995) Molecular genetics of plant embryogenesis. *Annu. Rev. Plant Physiol. Plant Mol. Biol.* **46**, 369–394.

Meinke, D.W. and Sussex, I.M. (1979) Embryo-lethal mutants of *Arabidopsis thaliana*: a model system for genetic analysis of plant embryo development. *Dev. Biol.* **72**, 50–61.

Meinke, D.W., Franzmann, L.H., Nickle, T.C. and Yeung, E.C. (1994) Leafy cotyledon mutants of *Arabidopsis. Plant Cell* **6**, 1049–1064.

Meyerowitz, E.M. and Somerville, C.R., eds (1994) *Arabidopsis*. Cold Spring Harbor Laboratory Press, Cold Spring Harbor.

Miséra, S., Müller, A.J., Weiland-Heidecker, U. and Jürgens, G. (1994) The *FUSCA* genes of *Arabidopsis*: negative regulators of light responses. *Mol. Gen. Genet.* **244**, 242–252.

Müller, A.J. (1963) Embryonentest zum Nachweis rezessiver Letalfaktoren bei *Arabidopsis thaliana. Biol. Zentralbl.* **82**, 133–163.

Neuffer, M.G., Coe, E.H. and Wessler, S.R., eds (1995) *Mutants of Maize*. Cold Spring Harbor Laboratory Press, Cold Spring Harbor (in press).

Patton, D.A. and Meinke, D.W. (1990) Ultrastructure of arrested embryos from lethal mutants of *Arabidopsis thaliana. Am. J. Bot.* **77**, 653–661.

Patton, D.A., Franzmann, L.H. and Meinke, D.W. (1991) Mapping genes essential for embryo development in *Arabidopsis thaliana. Mol. Gen. Genet.* **227**, 337–347.

Pepper, A., Delaney, T., Washburn, T., Poole, D. and Chory, J. (1994) *DET1*, a negative regulator of light-mediated development and gene expression in *Arabidopsis*, encodes a novel nuclear-localized protein. *Cell* **78**, 109–116.

Schneider, T., Dinkins, R., Robinson, K., Shellhammer, J. and Meinke, D.W. (1989) An embryo-lethal mutant of *Arabidopsis thaliana* is a biotin auxotroph. *Dev. Biol.* **131**, 161–167.

Schwartz, B.W., Yeung, E.C. and Meinke, D.W. (1994) Disruption of morphogenesis and transformation of the suspensor in abnormal suspensor mutants of *Arabidopsis. Development* **120**, 3235–3245.

Shellhammer, A.J. (1991) Analysis of a biotin auxotroph of *Arabidopsis thaliana*. Ph.D. Thesis. Oklahoma State University, Stillwater.

Shellhammer, J. and Meinke, D.W. (1990) Arrested embryos from the *bio1* auxotroph of *Arabidopsis* contain reduced levels of biotin. *Plant Physiol.* **93**, 1162–1167.

Shevell, D.E., Leu, W.-M., Gillmor, C.S., Xia, G., Feldmann, K.A. and Chua, N.-H. (1994) *EMB30* is essential for normal cell division, cell expansion, and cell adhesion in *Arabidopsis* and encodes a protein that has similarity to *Sec7. Cell* **77**, 1051–1062.

Vernon, D.M. and Meinke, D.W. (1994) Embryogenic transformation of the suspensor in *twin*, a polyembryonic mutant of *Arabidopsis. Dev. Biol.* **165**, 566–573.

Vernon, D.M. and Meinke, D.W. (1995) Late *embryo-defective* mutants of *Arabidopsis. Dev. Genet.* **16**, 311–320.

Wei, N., Chamovitz, D.A. and Deng, X.-W. (1994) *Arabidopsis COP9* is a component of a novel signaling complex mediating light control of development. *Cell* **78**, 117–124.

West, M.A.L. and Harada, J.J. (1993) Embryogenesis in higher plants: an overview. *Plant Cell* **5**, 1361–1369.

West, M.A.L., Yee, K.M., Danao, J., Zimmerman, J.L., Fischer, R.L., Goldberg, R.B. and Harada, J.J. (1994) *LEAFY COTYLEDON 1* is an essential regulator of late embryogenesis and cotyledon identity in *Arabidopsis*. *Plant Cell* **6**, 1731–1745.

Yadegari, R., de Paiva, G.R., Laux, T., Koltunow, A.M., Apuya, N., Zimmerman, J.L., Fischer, R.L., Harada, J.J. and Goldberg, R.B. (1994) Cell differentiation and morphogenesis are uncoupled in *Arabidopsis raspberry* embryos. *Plant Cell* **6**, 1713–1729.

Yeung, E.C. and Meinke, D.W. (1993) Embryogenesis in angiosperms: development of the suspensor. *Plant Cell* **5**, 1371–1381.

Insertional mutagenesis to dissect embryonic development in *Arabidopsis*

K. Lindsey, J.F. Topping, P.S.C.F. da Rocha, K.L. Horne, P.R. Muskett, V.J. May and W. Wei

4.1 Introduction

In contrast to the situation in animals, the establishment of organ systems in plants is primarily a post-embryonic process: plant embryos are not miniature versions of the adult. Nevertheless, events occur during plant embryogenesis that are critical for generating the body plan of the seedling, and it is during this phase of the life cycle that the shoot and root meristems are formed which, perhaps more than any other single feature of plants, distinguish their growth habit from that of members of the animal kingdom.

There are also some dramatic differences between monocotyledonous (monocot) and dicotyledonous (dicot) species with regard to the pattern of embryonic development, and these have been reviewed elsewhere (Lindsey and Topping, 1993; West and Harada, 1993). For a detailed consideration of monocot embryogenesis, see Chapter 6, Clark. An excellent model system for the study of the genetics of early development in dicots is represented by embryogenesis in *Arabidopsis* which is, in morphological terms, relatively simple and very well defined (Meinke and Sussex, 1979; Meinke *et al.*, 1989). While this species has many advantageous features that favour its use as a model, it is by no means typical of dicots in the precise patterns of cell divisions that occur, since there is much diversity in detail between species. Nevertheless, a generalized cascade of events takes place that is shared by many dicot species. During the 12 to 14 days over which seed development occurs in *Arabidopsis*, the embryo undergoes a range of structural and biochemical changes that create a highly patterned multicellular individual, in which the fate of particular cells can be predicted (Furner and Pumfrey, 1992; Irish and Sussex, 1992; Scheres *et al.*, 1994), and which becomes desiccated in readiness to exploit the environmental conditions that will favour germination and precipitate new growth. The salient features of embryonic morphogenesis in *Arabidopsis* are described in detail (see Chapter 2, Torres-Ruiz *et al.* and Chapter 3, Meinke), and will not be repeated here. However, it is relevant to emphasize that, in a number of ways, the heart stage of embryogenesis represents a significant point in the development of the *Arabidopsis* plant. By this stage, precursors of the three fundamental tissues of the seedling, namely the epidermis, ground tissue (comprising the cortex and root endodermis) and vascular tissue (i.e. all cells internal to the endodermis), have been laid down, and it is possible to identify with some accuracy (particularly for root progenitor cells) the fates of many of the component cells. This fate mapping of cells in the embryonic shoot and root meristems has been achieved by clonal analysis, in which marked sectors have been generated either by mutagenesis or by transposon-activated reporter gene activity in cell lineages (Furner and Pumfrey, 1992; Irish and Sussex, 1992; Scheres *et al.*, 1994). Furthermore, it is during the heart stage that the first evidence of root meristem activity is detectable, with the appearance of initial cells that contribute to distinct cell files in the seedling root (Dolan *et al.*, 1994; Scheres *et al.*, 1994). Later stages of embryo morphogenesis in *Arabidopsis* are essentially devoted to an elaboration of the structures established by the heart stage. In contrast to the root, cell fate in the shoot meristem of the mature embryo is much less predictable, such that chance events in division at the apex can be influential, and it is suggested that this may be a reflection of the small size of the embryonic shoot apex of *Arabidopsis* (Furner and Pumfrey, 1992).

A further important feature of the older embryo is the elaboration of a biochemical differentiation of embryonic tissues. Chlorophyll accumulation is first detectable during the transition between heart stage and torpedo stage but, following desiccation, the embryo of the mature seed is yellow in colour. While the endosperm represents a nutritive tissue in the younger seed, high-energy metabolites, notably lipids and starch, accumulate in the maturing embryo, to be mobilized during germination. Dramatic changes occur in the profile of accumulated embryonic proteins. The latter part of embryogenesis in many species, including *Arabidopsis*, is characterized by the synthesis and accumulation of three categories of major proteins: the storage proteins (in *Arabidopsis*, for example, the 2S and 12S proteins), which represent a supply of amino acids that are available to the germinating seedling (Goldberg *et al.*, 1989; Thomas, 1993); the so-called 'late-embryogenesis-abundant' proteins, which may play a role in protection of the seed against the effects of desiccation (Dure *et al.*, 1989); and diverse enzymes that will be required for the mobilization of the stored food reserves (Thomas, 1993).

The regulation of each of these processes — morphogenesis, biochemical differentiation and desiccation — and their co-ordination to produce a fully integrated developmental pathway, is clearly physiologically and genetically complex, and it is not the aim of this chapter to discuss all aspects. For an excellent overview of seed biology, the reader is directed to Bewley and Black (1994).

Instead, we shall focus our attention on genetic mechanisms involved in *Arabidopsis* embryo development, and describe our approach for analysis of the relationship between morphogenesis and cytodifferentiation in the embryo and seedling. In particular, we shall describe a strategy of insertional mutagenesis and promoter trapping, established in this laboratory, that allows a genetic and cellular dissection of embryonic development.

4.2 The genetic approach to the study of embryonic development

Development in multicellular organisms can be considered to consist of two fundamental components: the establishment of correct shape (pattern formation and morphogenesis), and the creation of a division of labour between component cells (cytodifferentiation). The term *pattern formation* has been used in slightly different ways in different developmental contexts: for example, in describing the spatial organization of floral organs (Coen and Meyerowitz, 1991) or the relative positioning of individual cells within a tissue, such as the root epidermis (e.g. Dolan *et al.*, 1994). A common denominator in these interpretations is the concept of relative position between independent components (be they cells or organs) in a system (a tissue or collection of organs), regardless of their individual identities. It might be considered reasonable to use this term to describe the processes whereby cells in the developing embryo become organized in the correct spatial arrangement. However, as we shall see, this term may be inadequate to account for the processes that determine the overall shape of the embryo, since evidence will be presented that components of pattern formation may occur in the absence of correct spatial patterning of individual cells. In this article, therefore, we shall use the term *morphogenesis* to describe the generation of *shape,* the term *pattern formation* to describe the processes whereby the relative positions of organs or parts of the embryo become spatially organized, relative to one another, and the term *cytodifferentiation* to describe changes in the biochemical and structural properties of individual cells which lead to functional specialization.

Essential components of the establishment of the body plan have been discussed in a series of publications from the laboratory of Jürgens (e.g. Berleth and Jürgens, 1993; Mayer *et al.*, 1991, 1993; Torres-Ruiz and Jürgens, 1994). These facets of pattern formation, established during embryogenesis and evident in the germinated seedling, are discussed in detail (see Chapter 2, Torres-Ruiz *et al.*). *Apical-basal polarity* is established during the very earliest stages of embryogenesis, beginning with the asymmetrical division of the zygote, which in turn reflects the polarity evident in the egg cell and zygote itself (Reiser and Fischer, 1993), presumably due, to a significant extent, to the effects of maternal factors. This polarity is apparent in the seedling and, of course, throughout the life of the plant, in the differences in cellular organization and developmental fate of the shoot and root apical meristems, respectively. *Radial symmetry*, as occurs in the hypocotyl, root and stem of the plant, is first seen during the development of the proembryo. *Bilateral symmetry*, established at the heart stage of embryogenesis, is also maintained in the seedling, with its pair of cotyledons. Therefore, during the first 3 or 4 days of its development, the embryo sets up what might be considered to be 'cellular domains', which are regions of predictable fate, namely the apical meristems, the ground tissues and the vascular tissues, and subsequent development merely elaborates the size and shape of these domains, building upon the cellular blueprint that has been drawn up. In this sense, plant development can be considered to be 'modular'.

Cytodifferentiation in the *Arabidopsis* embryo, as we have seen, is relatively limited, such that the mature embryo contains few cell types, although the fate of the cells, at least in the embryonic root, can be mapped with some precision. It is a dogma of developmental biology that the generation of cell diversity is a process 'downstream' of pattern/shape formation, such that positional information or, possibly, events established early on and transmitted down cell lineages, determine the final fate of a cell. Certainly in plant biology there is evidence from the analysis of chimeras and from tissue-culture experiments to support the view that cells have no, or only limited, developmental 'memory', and differentiate according to where they are situated, and thus according to the molecular signals they receive, within a three-dimensional organ (e.g. Becraft and Freeling, 1991; Dawe and Freeling, 1991).

What mechanisms determine the establishment of these facets of development? How do embryos make decisions about the positioning of the cellular domains that constitute the body plan, and how do cells, within those domains, differentiate correctly? In general terms, it might be expected that the establishment of, for example, the embryonic shoot or root apical meristems would involve the localized activity of specific regulatory proteins, presumably as a consequence of the localized expression of particular genes. It is known, for example, that the identity of floral organs in *Arabidopsis* and other species is regulated by the combined action of genes encoding transcription factors, each of which is expressed spatially in a very restricted fashion (see e.g. Coen and Meyerowitz, 1991; Okamuro *et al.*, 1993). There is also evidence that the identity of individual cell types within an organ may be dependent upon the cell-autonomous expression of regulatory molecules. Furthermore, in some cases at least, transcription factors, required for example in trichome differentiation (Larkin *et al.*, 1993), may be involved. This information was obtained following a genetic approach to dissect events in development, a strategy used to great advantage in other model experimental organisms such as *Drosophila* (e.g. Hoch and Jäckle, 1993; Kornberg and Tabata, 1993), *Caenorhabditis* (Priess, 1994) and zebrafish (Mullins and Nüsslein-Volhard, 1993). By using a mutational approach, an enormous amount of information can be obtained about the roles of specific genes in particular facets of a developmental pathway, and about the interactions between the genes identified, even in the absence of isolated genes. Since the late 1970s, particularly in the laboratory of Meinke, a chemical mutagenesis strategy has been used to identify mutants whose embryonic development and, especially, viability is affected (Meinke, 1986; Meinke and Sussex, 1979). In more recent years, many insights into embryonic pattern formation and morphogenesis have been derived from the same experimental strategy, and some of the key features of relevance to our own work will be summarized here.

Table 4.1 shows a list of genes, identified by chemical (EMS) mutagenesis or T-DNA tagging screens, that affect aspects of (a) embryonic pattern or shape formation and (b) embryonic cytodifferentiation. Mayer *et al.* (1991), for example, identified nine loci that are required for the correct establishment of embryonic

Table 4.1. *Some mutants affecting embryonic development in* Arabidopsis[a]

Mutant	Phenotype	Reference
(a) Pattern/shape		
gnom/emb30	Apical-basal deletion	Mayer *et al.*, 1991[E], 1993[E]; Shevell *et al.*, 1994[T]
gurke	Apical deletion	Mayer *et al.*, 1991[E]
monopteros	Basal deletion	Mayer *et al.*, 1991[E]; Berleth and Jürgens, 1993[E]
fackel	Central deletion	Mayer *et al.*, 1991[E]
knolle	Radial pattern defect	Mayer *et al.*, 1991[E]
keule	Radial pattern defect	Mayer *et al.*, 1991[E]
fass	Shape defect	Mayer *et al.*, 1991[E]; Torres-Ruiz and Jürgens, 1994[E]
pin1	Shape defect	Liu *et al.*, 1993[E]
mickey	Shape defect	Mayer *et al.*, 1991[E]
knopf	Shape defect	Mayer *et al.*, 1991[E]
amp1	Shape defect	Chaudhury *et al.*, 1993[E]
hydra	Shape defect	Topping, May and Lindsey[T], unpublished data
bashful	Shape defect	da Rocha and Lindsey[T], unpublished data
(b) Differentiation		
fusca	Pigmentation	Miséra *et al.*, 1994[E]; Castle and Meinke, 1994[T]; Bäumlein *et al.*, 1994[E]; Keith *et al.*, 1994[E]
suspensor	Enlarged suspensor	Schwartz *et al.*, 1994[T]; Horne and Lindsey[T], unpublished data
leafy cotyledon	Cotyledon identity	Meinke, 1992[T]; West *et al.*, 1994[T]
raspberry	Enlarged suspensor	Yadegari *et al.*, 1994
albino	Pigmentation	Meinke, 1986[E]; Wei and Lindsey[T], unpublished data

[a]A large number of embryo-defective mutations have now been mapped (Franzmann *et al.*, 1995). Mutants found in populations generated by ethyl methane sulphonate (EMS) are indicated by superscript[E], and mutants found in T-DNA screens are designated by superscript[T].

pattern and shape. This mutational analysis has furthermore provided strong evidence that the development of the embryo can indeed be considered to be modular, since mutations in specific single genes can completely inhibit the development of individual cellular domains, without affecting others. For example, the *monopteros* mutation (Berleth and Jürgens, 1993; Mayer *et al.*, 1991) results in the failure of the hypocotyl and root to develop, although the shoot meristem and cotyledons remain unaffected. The *gnom* mutation, in contrast, results in elimination of the apical meristems (Mayer *et al.*, 1991, 1993), and it is tempting to suggest that this is a reflection of the fact that, in this mutant, the first detectable lesion in the embryo is the aberrant symmetrical rather than the typically asymmetrical first division of the zygote. The *GNOM* gene, also designated *EMB30*, has now been cloned (it is the only cloned gene known to affect embryonic pattern formation; Shevell *et al.*, 1994), and its possible role in the regulation of embryo shape will be discussed later.

A number of genes affect embryonic cell differentiation, but without appearing primarily to affect pattern formation or shape of the embryo. The *sus* mutants (see Chapter 3, Meinke; Schwartz *et al.*, 1994) and *raspberry* mutants (Yadegari *et al.*, 1994), for example, which display defects in the development of the suspensor, show evidence of biochemical differentiation of the embryo proper in the absence of appropriate structural differentiation. This will be discussed in more detail later.

The majority of the genes listed in Table 4.1 were identified following chemical mutagenesis screens, and while the mutants themselves provide a large amount of information about developmental processes in embryogenesis; even more information would be obtained if the respective genes could be cloned. For example, their cellular expression patterns could be determined by *in situ* hybridization analysis, the effects of ectopic expression of the genes could be analysed, and proteins that interact with and regulate the promoters of the genes, to modulate their expression during development, could be investigated.

For EMS-generated mutants, gene isolation is most commonly attempted by positional or map-based cloning. This approach has proved successful in recent years (e.g. Giraudat *et al.*, 1992; Leyser *et al.*, 1993), but can be rather time-consuming. A more direct approach for cloning genes is available if mutants can be obtained by insertional mutagenesis, either using endogenous (Tsay *et al.*, 1993) or heterologous transposons (Aarts *et al.*, 1993; Bancroft *et al.*, 1993), or T-DNAs (Feldmann, 1991; Topping and Lindsey, 1995; Walden *et al.*, 1991). The advantage here is that well-characterized transposon or T-DNA sequences, physically linked to a mutant gene, can be exploited to isolate the tagged sequence by such methods as plasmid rescue or inverse PCR. The isolated plant DNA can then be used as a molecular probe to rescue a wild-type allele of the gene from a cDNA or genomic library.

One disadvantage of an insertional mutagenesis approach, in the context of genetic dissection of a developmental pathway, is the number of transformants which must be generated and screened in order to identify a range of genes and alleles that regulate a given process. This is potentially less of a problem if transposons are used, as long as they are sited close to target genes of interest (which must be mapped). Targeted transposon mutagenesis, using introduced maize element systems such as Ac/Ds or En/I, will certainly emerge in the next few years as an important technique, but at present the frequency of transposition of these elements in *Arabidopsis* is too low to identify large numbers of new genes. Similarly, while T-DNA insertional mutagenesis has proved very powerful for the identification and cloning of genes, it is currently of limited impact in programmes of saturation mutagenesis (although with rapid advances being made in the mass production of transgenic *Arabidopsis*, this situation may change). Nevertheless, we consider that the insertional mutagenesis approach is particularly useful if used in parallel with EMS screens. These can reveal the types of phenotype of interest, and may be used to identify multiple alleles of specific genes, while tagging strategies can be exploited to clone single alleles. We shall now discuss our own approach of insertional mutagenesis for analysis of embryonic development in *Arabidopsis*.

4.3 Insertional mutagenesis and promoter trapping

Our experimental strategy to investigate the relationship between embryonic morphogenesis and cytodifferentiation has two main objectives: (a) to generate mutants that display defects in embryo shape; and (b) to generate markers of specific cell types, in order to characterize cell fate in the mutants. We have achieved this by the design of a T-DNA which can not only function as an insertional mutagen, but is also capable of acting as a functional tag of mutant gene promoters. The structure of this 'promoter trap' T-DNA, in a Bin19-based vector designated pΔgusBin19 (Topping *et al.*, 1991), is presented in Figure 4.1. The full sequence of the T-DNA has been determined and is available (Wei *et al.*, 1994b). At the right border of the T-DNA is a constitutively expressed selectable marker, the *NPT-II* gene, that confers resistance to kanamycin in transgenic plants. At the left border is the 'promoter trap'. This comprises a promoter-less *gus*A gene with its 5′ end flanked by the T-DNA border sequence. The rationale of this design is that the *gus* gene may be activated following insertion of the T-DNA downstream of a native gene promoter, and in this situation a functional fusion may be generated. Since the T-DNA left-border region contains translational stop codons in all reading frames, it is most likely that a transcriptional fusion will be generated. The *gus* coding region has its own ATG codon. We describe this gene tag as an 'interposon'. Activation of the *gus* gene by a native gene promoter is expected to result in a spatial and temporal GUS activity pattern in transgenic tissues that is similar if not identical to the expression pattern of the tagged gene. Using this vector to generate a population of transgenics, the aim was to screen seedlings for (a) segregating mutant phenotypes, in which embryonic development was altered, and (b) cell-type-specific GUS activity in individual transgenic lines, which would provide both visual markers for cell-fate analysis in mutants, and simultaneously facilitate the cloning of the GUS fusion gene, if this was of interest.

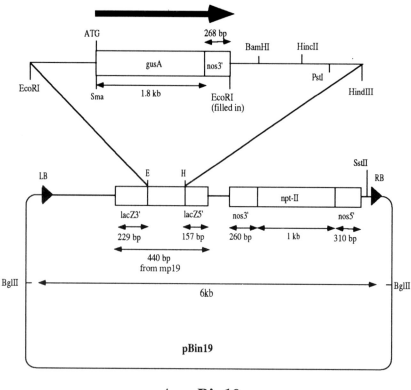

pΔgusBin19
12 kb

Figure 4.1. *The promoter trap plasmid pΔgusBin19.*

In establishing this gene-tagging strategy, one important question concerns the frequency at which phenotypic mutants and active GUS fusions could be detected using pΔgusBin19. The aim of early work was to improve the frequency of methods for *Agrobacterium*-mediated transformation, such that we were able to generate over 2000 independent transgenic plants for screening purposes (Clarke *et al.*, 1992). A number of different classes of phenotypic mutant have been identified in this population, although at the time of writing only about 40–50% of the transformants produced have been bulked up and screened for mutants. Of significance to an analysis of embryonic development, we have identified mutants that are defective in both pattern formation and shape (apical-basal polarity, radial symmetry, bilateral symmetry) and in embryonic cytodifferentiation (albino, *fusca*, aberrant suspensor). The overall frequency of embryonic mutants in this population is difficult to determine with accuracy, because of the small sample size. However, we have found seven independent mutants that display defects in seedling shape and are also known to be altered in embryogenesis, in a screen of approximately 700 transformed lines, which represents about 1%. In the same screen we have also identified two *fusca* mutants, one suspensor mutant, three albinos and two embryonic lethals of uncharacterized phenotype. Two other seedling shape mutants have also been identified, but we have yet to determine whether the embryos are also abnormal. We shall discuss some of these mutants below in more detail.

The frequency of active GUS fusions generated in the *Arabidopsis* populations transgenic for pΔgusBin19 is relatively high (Lindsey *et al.*, 1993; Topping *et al.*, 1991, 1994). For example, approximately 30% of the transformants analysed exhibited GUS activity in their roots; a similar proportion showed activity in leaf tissues, while about 20% were found to display activity in floral organs and siliques, respectively (Lindsey *et al.*, 1993). This frequency is essentially independent of genome size, such that a similarly high frequency of Δ*gus* activation occurs in tobacco (Topping *et al.*, 1991), the genome of which is approximately 50-fold larger than that of *Arabidopsis* and contains 60% repetitive DNA sequences, compared with only 20% in *Arabidopsis*. These data are consistent with the results of other similar studies in which the activation of promoter-less reporter genes was analysed in the two species (Kertbundit *et al.*, 1991; Koncz *et al.*, 1989).

Histochemical analyses of promoter-less *gus* activation patterns have revealed GUS enzyme activities in a wide range of tissues, following transformation with different interposon vectors (Kertbundit *et al.*, 1991; Lindsey *et al.*, 1993). For example, within roots of *Arabidopsis* we have identified GUS activity restricted to individual cell layers (epidermis, cortex, pericycle and columella; unpublished data) as well as within several cell types of the root vascular bundle, either including or excluding the endodermis, and in the root meristem (unpublished data).

The promoter-trapping system described above therefore provides diverse cell-type-specific markers that can be of value in analysing the cellular organization of mutants. At the same time, the gene fusions generated *in vivo* allow facilitated cloning of the tagged genes, and phenotypic mutants can be generated. Before describing in any detail the analysis of mutants, we shall summarize the results of our progress in cloning embryonically expressed genes, based on the identification of GUS fusion activity rather than on mutant phenotype.

4.3.1 *Cloning fusion genes expressed in embryos*

In order to identify transgenic lines in which embryonically expressed genes had been tagged, transgenic plants were screened in the first instance for GUS activity in siliques, using the very sensitive fluorimetric assay (Jefferson *et al.*, 1987). Siliques containing seeds with embryos ranging from the globular stage to the cotyledonary stage (up to 10 days post-fertilization) were removed from each of 430 independent primary transformants (T1 plants) containing pΔgusBin19, and assayed. A total of 74 of the 430 lines analysed (17.2 %) were found to be GUS positive (Topping *et al.*, 1994). In order to determine which of these lines exhibited GUS activity within embryos, T1 seed (which contains T2 embryos) was germinated and T2 plants were bulked up for further analysis. Of the 74 GUS-positive lines, 30 lines that displayed the highest levels of GUS activity were analysed in more detail by GUS histochemistry. Of the 30 GUS-positive lines, 6 exhibited GUS activity in the embryo, which represents 20% of the GUS-positive siliques analysed. This indicates an overall frequency of 3–4% GUS fusion activity in embryos. One line was found to exhibit expression that was specific to the embryo. GUS fusion activity in three lines that were demonstrated by Southern blot/hybridization analysis to contain single T-DNA inserts was investigated in more detail. These lines were designated AtEM101, AtEM201 and AtEN101, respectively.

GUS activity in the seeds of line AtEM101 was found to be restricted to the embryo. No activity was detectable in globular stage embryos (2–3 days post-fertilization), but during the heart stage (3–4 days post-fertilization) activity was detected in the basal part of the embryo (Figure 4.2a). No activity was found in the suspensor or in the upper part of the embryo. During the subsequent stages of embryogenesis, GUS activity was found to be restricted principally to the basal part of the embryo, constituting the developing root primordium. On germination, the highest levels of GUS activity were observed in the root tips.

GUS fusion activity in seeds of AtEM201 was found to be restricted to the embryo, as in the case of AtEM101, but the spatial and temporal pattern was different. GUS activity was first detectable much earlier in embryogenesis, in all cells of the 8-celled embryo proper and subsequently in globular embryos (2 days post-fertilization; Figure 4.2b). In contrast to the situation in AtEM101, GUS activity was located throughout the entire heart-stage embryo. No activity was observed in the suspensor at any stage. As embryogenesis progressed, activity was found to reach relatively high levels in the developing radicle, with faint staining activity in the cotyledons. In germinated seedlings, GUS activity was detected at low levels in the cotyledons, but higher levels were found in the shoot and root apices. In mature flowering plants, activity was restricted to the root tips, including the tips of lateral roots.

A third pattern of GUS activity is exhibited by line AtEN101. GUS expression was activated early on, but was not restricted to the embryo, as it was also found in the endosperm. Activity was transient in both embryo and endosperm at the early globular and late globular/early heart stages (Figure 4.2c), and no activity was detected in the suspensor. Isolation of embryos followed by incubation in 5-bromo-4-chloro-3-indolyl glucuronide (X-gluc) revealed that, by late heart stage, embryos exhibited no GUS activity. In germinated seedlings, low levels of activity were detectable in the cotyledons, the shoot apex and the root elongation zone and root cap, but not in the root meristem. In mature flowering plants, the leaves and roots show no GUS activity, which is restricted to the tapetum and stigmatal papillae.

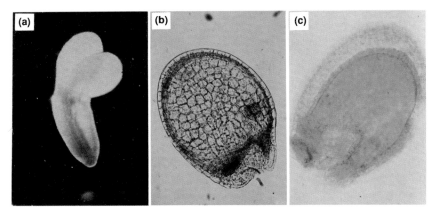

Figure 4.2. *GUS fusion activity in three transgenic lines of* Arabidopsis *transformed with pΔgusBin19: (a) AtEM101; (b) AtEM201; (c) AtEN101.*

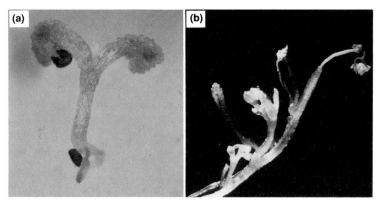

Figure 4.4. fusca *mutant: (a) light-grown seedling; (b) dark-grown seedling showing flowers.*

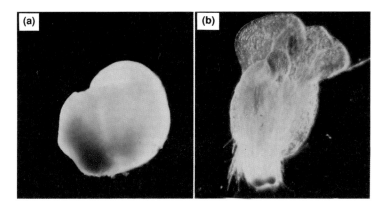

Figure 4.7. hydra *mutant containing the AtEM101 root meristem marker: (a)* hydra *embryo; (b)* hydra *seedling.*

4.3.2 *Fusion transcripts*

One question that arises is whether the mRNAs transcribed from the *gus*A sequences indeed represent fusions with native plant genes. A recent paper (Fobert *et al.*, 1994) describes how a promoter-less *gus* gene, introduced into tobacco, was activated by a suspected cryptic promoter which was not associated with a gene. Northern blots carried out on lines AtEM101, AtEM201 and AtEN101 demonstrated that T-DNA sequences did not act as cryptic promoters. The results showed that single *gus*A fusion transcripts in the approximate range 2.8–4.0 kb were generated in each transgenic line, respectively, and the relative transcript abundances correlated broadly with the GUS activity in seedlings of the three respective transgenic lines (Topping *et al.*, 1994, and unpublished data). Since the maximum expected T-DNA-derived transcript is 2.5 kb (1.8 kb *gus*A plus 0.7 kb T-DNA left border; Wei *et al.*, 1994b), this result demonstrates that transcription in each line was initiated up to approximately 1.5 kb into the plant genomic DNA flanking the T-DNA left-border sequence.

In order to investigate whether the sequences into which the T-DNAs had inserted represent transcribed genes, the flanking sequences at the T-DNA left borders in all three lines were amplified by inverse polymerase chain reaction (IPCR), using primers based on T-DNA sequences (Topping *et al.*, 1994). The IPCR products were cloned and sequenced, and used as probes of cDNA and genomic libraries of non-transformed *Arabidopsis* plants. Sequencing of the IPCR products allowed characterization of the T-DNA plant junctions at the left border, and these are illustrated in Figure 4.3. For lines AtEM201 and AtEN101, cDNAs corresponding to the tagged sequences were identified, demonstrating that in each case a transcribed gene had been disrupted by the T-DNA (Topping *et al.*, 1994, and unpublished data). In the case of AtEM101, we currently believe that the T-DNA has integrated into an intron, precluding the use of the IPCR products as a probe for a cDNA. Nevertheless, for this tagged sequence, *in situ* hybridization experiments have been carried out using cloned genomic sequences in order to determine whether a corresponding native gene transcript is localized in the seed. Our results demonstrate that the probe hybridizes to a transcript in the wild-type

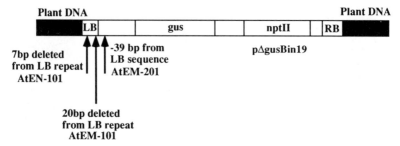

Figure 4.3. *Sites of T-DNA/plant DNA junctions in transgenic lines AtEM101, AtEM201 and AtEN101.*

embryo, which parallels the localization of the *gus* transcript (J. Nielsen, J. Topping and K. Lindsey, unpublished data).

What do the tagged genes encode? At the time of writing, *AtEM101* cDNA isolation is in progress, and the *AtEM201* cDNA has not yet been completely sequenced. On the basis of the observed expression patterns, we might expect the genes to be involved in meristematic cell activity. However, the tagged *AtEN101* gene has been found to encode a 3-β-ketoacyl-CoA thiolase. Sequence analysis has revealed a complex situation in line AtEN101. The T-DNA has been shown to have integrated in the first intron of the thiolase gene, which encodes two alternative transcripts, but the latter and the reporter gene *gus*A are oriented in a divergent manner. Therefore, although the thiolase transcripts would be expected to be expressed during seed development, the expression patterns will be somewhat different to the GUS fusion activity, and this is currently being investigated. The 5' end of the fusion transcript is also being characterized.

4.4 Characterization of phenotypic mutants

Interestingly, a number of the transgenic lines that exhibit GUS fusion activity do not display obvious phenotypic abnormalities. There are at least three possible reasons for this. Firstly, the mutant gene may be functionally redundant, its role being fully or partially complemented by a different gene. Secondly, if a T-DNA inserts into an intron, it may be removed during RNA processing; we have identified insertion events in introns that fail to generate clear phenotypic mutations. Thirdly, a phenotypic mutant may be conditional, only being revealed under certain environmental conditions. An excellent illustration of this possibility is provided by mutants that are defective in phytochrome A. When grown under standard greenhouse conditions (i.e. in white light) the mutants are phenotypically normal, and they only exhibit growth defects when grown in light enriched with far-red wavelengths (Whitelam *et al.*, 1993). It is therefore clearly crucial, in mutagenesis studies, to design appropriate screens for particular types of mutation. The generation of GUS fusions in particular tissues or organs is a useful aid in focusing the search for specific classes of mutant, and indeed we have identified subtle phenotypic aberrations in several transgenic lines originally studied on the basis of their GUS fusion activity. A number of mutants that display defective root development fall within this category (unpublished data).

Our search for mutants that exhibit defective embryonic morphology has, however, primarily been carried out by screening for seedling shape mutants shortly after seed germination, a strategy described and used with dramatic success by Mayer *et al.* (1991). The rationale is that defects that become apparent in young seedlings may first be recognized in the embryo, implicating that embryonic process in later development. We have identified mutants that display defective seedling shape, and we have also found mutants that exhibit abnormalities in aspects of embryonic cytodifferentiation, although embryonic pattern formation may proceed normally. We have carried out only very preliminary characterizations of this latter category of mutants, but we shall summarize the principal observations here.

4.4.1 *Embryonic differentiation mutants*

fusca mutants. Two *fusca* mutants were identified in our seedling screens, recognizable by both a lack of chlorophyll and a distinctive pink colour, due to inappropriate anthocyanin accumulation in the cotyledons and, to a lesser extent, in the hypocotyl (Figure 4.4). The mutants are lethal to seedlings. The mutant shown in Figure 4.4a not only accumulates anthocyanin, but is also characterized by a cotyledonary epidermis composed of abnormally vacuolated cells. A genetic analysis of the *fusca* phenotype by Miséra *et al.* (1994) has identified nine complementation groups. There is now genetic evidence that the *fusca* mutants are allelic to the so-called *c*onstitutively *p*hotomorphogenic (*cop*) mutants, which fail to exhibit the normal etiolation response (Miséra *et al.*, 1994; Wei *et al.*, 1994a). When grown in the dark, our mutant similarly undergoes continued morphogenesis, developing elongated cotyledonary petioles, leaves and even flowers (Figure 4.4b). This phenotype is similar to that of *cop*2 (Wei *et al.*, 1994a), but complementation tests have yet to be carried out.

Some *fusca* mutants are defective in other aspects of embryonic cell differentiation. For example, the *fus*3 mutant is defective in the accumulation of storage proteins and lipids and the control of organ identity (the cotyledons possess trichomes, normally only a characteristic of leaves), as well as other aspects of seed physiology such as establishment of dormancy and dessication tolerance (Bäumlein *et al.*, 1994; Keith *et al.*, 1994). *fus*6 mutants exhibit different abnormalities of embryonic cell differentiation; the accumulation of lipids and protein bodies appears to be normal, but the subcellular structure of the cells in the embryonic shoot apex is aberrant, being highly vacuolated, while these cells are densely cytoplasmic in wild-type embryos. This correlates with defective shoot development in the germinated seedling.

The *FUS6* gene has been tagged by a T-DNA and cloned, and encodes a novel class of protein with no significant homology to other known proteins, but possessing a number of putative protein kinase C phosphorylation sites (Castle and Meinke, 1994). The *FUS1/COP1* and *FUS7/COP9* genes have also been cloned, the former encoding a protein with putative G-protein and DNA-binding domains (Deng *et al.*, 1992), and the latter being a protein distinct from either FUS1 or FUS6 (Wei and Deng, 1993). It is suggested that the FUSCA-type proteins may be involved, perhaps in concert with each other, in intracellular signalling mechanisms, mediating downstream growth responses.

suspensor mutants. We have identified one mutant that displays a defect in the differentiation of the basal cell, the larger product of the first zygotic division (and the progenitor cell of the suspensor; Yeung and Meinke, 1993). In the wild-type seed, the suspensor undergoes apoptosis, and by the cotyledonary stage of embryogenesis it has disappeared. However, in this mutant the suspensor divides excessively, while development of the embryo proper is restricted to a globular structure, although embryonic cellular spatial organization is not dramatically disturbed (Figure 4.5). A number of so-called *suspensor* (*sus*) mutants with broadly similar phenotypes have been described by Schwartz *et al.* (1994), and the *raspberry* mutants described by Yadegari *et al.* (1994) exhibit similar abnormalities. Crossing experiments with *sus*1, which is phenotypically most similar to our mutant, indicates that the genes are non-allelic and a new gene affecting suspensor fate has been defined.

In a number of the *suspensor* mutants, the embryos may show some structural abnormalities, for example in the organization of cells at the base of the embryo (*sus1-1*) or in the outer cell layers (*sus2-1*, *sus3*, *raspberry1*, *raspberry2*). Of significant interest, however, is the observation that the suspensor cells and cells of the embryo proper show evidence of abnormal biochemical differentiation, in the accumulation of storage proteins, lipids and starch grains, which in wild-type seeds are restricted to the cotyledons of maturing embryos. This indicates that at least some aspects of cell differentiation can take place during development, in the absence of correct pattern formation.

4.4.2 *Mutants displaying defects in shape*

We have identified a number of mutants that exhibit defects in seedling shape, and in which these defects are thought to originate during embryogenesis. Some are better characterized than others, and we shall now summarize the relevant data.

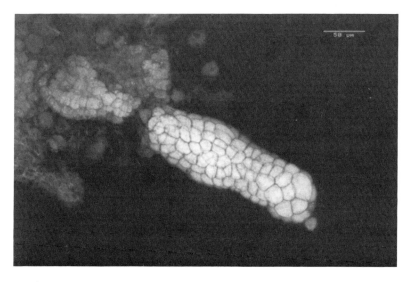

Figure 4.5. suspensor *mutant.*

bashful *mutant*. This was originally identified in a seedling screen as a dwarfed and slow-growing phenotype (Figure 4.6a). The leaves, roots and flowers of the homozygote are small, and within the siliques of the mutant are embryos that exhibit a range of abnormal shapes. In plants heterozygous for the mutant gene, however, no embryos with dramatically altered phenotypes are found, and we

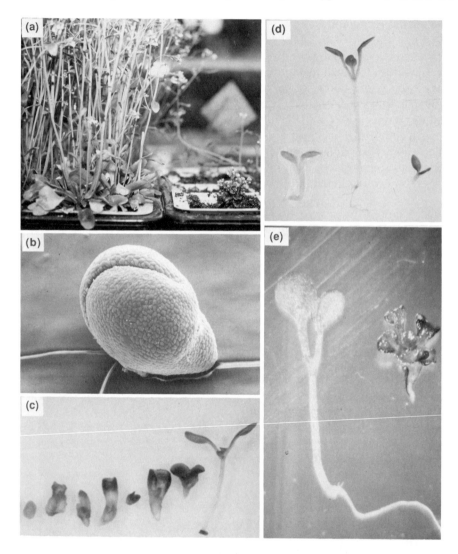

Figure 4.6. *Embryonic/seedling shape mutants: (a)* bashful *seedling on right, wild type on left; (b)* emb30 *embryo; (c)* emb30 *seedlings, showing inter-sibling variation, wild-type seedling on far right; (d)* rootless *seedling on right, wild type in centre, kanamycin-sensitive wild type on left; (e)* hydra *seedling on right, wild type on left.*

conclude that the effect of the mutant gene is a maternal one, which may be mediated by mechanical constraints on embryonic morphogenesis, brought about by spatial restrictions within the dwarfed silique. The primary transgenic line contained two segregating T-DNA copies, and the mutant contains a T-DNA that co-segregates with the mutant phenotype. A detailed molecular characterization of the mutant is in progress.

emb30 *mutant*. A second mutant, which we originally named *golftee* after the shape of at least some of the homozygous individuals, and which is tagged by a T-DNA, is now known to be allelic to the *emb30/gnom* mutant (Mayer *et al.*, 1993; see also Chapter 2, Torres-Ruiz *et al.* and Chapter 3, Meinke). The mutant seedling exhibits dramatic abnormalities in the organization of cells at the shoot and root apices, such that growth from these regions does not occur. Anatomical analysis of the embryo reveals that, in all alleles studied, the earliest divisions are abnormal, including the first division of the zygote; whereas this division is typically asymmetrical in wild-type plants, it tends towards symmetry in the *emb30* mutant. Further divisions of the enlarged apical cell are also abnormal, such that the octant and globular stage embryos have more cells than the wild-type. The lack of control over cell size or cell shape is also observed in older embryos (Figure 4.6b). Seedlings, even when they are siblings of a single mutant allele, interestingly display a wide range of phenotypes, from spherical to conical structures (Mayer *et al.*, 1993; Figure 4.6c), and when maintained on a simple culture medium (which supports the growth of wild-type plants) they fail to maintain structure, developing into callus. The EMB30 gene has been cloned following its disruption by T-DNA insertional mutagenesis, and it has been found to show homology to the yeast *SEC7* gene, encoding a protein involved in the secretory pathway (Shevell *et al.*, 1994). The role of the EMB30 gene is at present unknown, but it may be involved in secretory pathways essential for the control of cell growth or shape or, in view of the fact that inhibition of polar auxin transport in embryos can generate a similar embryonic phenotype to the *emb30* mutant (Liu *et al.*, 1993), it may play a role in this process.

GUS fusion activity from the promoter-less *gus* interposon in our mutant allele has generated *in vivo* GUS activity that, in the heterozygote, is specific to the pericycle of the root. However, in the mutant phenotype no GUS activity is detectable, indicating a lack of pericycle.

rootless *mutant*. This mutant is most clearly defective with regard to the development of the root system. The germinated seedling has phenotypically wild-type cotyledons, but an extremely reduced hypocotyl and root (Figure 4.6d). As the seedling develops, the leaves may occasionally show evidence of abnormal shape, especially in being narrow, and the roots form slowly. The overall architecture of the root system is abnormal, with a much reduced primary root and more lateral roots. The mutant can form flowers, but both flowers and siliques are small. One further feature of the *rootless* seedling is that the vascular connections between the cotyledons and the hypocotyl are often poorly formed, although the pattern of vascularization in the cotyledons themselves is not abnormal. At the time of writing we have not identified the point during embryogenesis when the developmental lesion first becomes apparent, but research is in progress.

One mutant that shows some similarities in seedling phenotype is *monopteros* (Berleth and Jürgens, 1993). The lack of root development in the seedling in this mutant is correlated with abnormal divisions in the basal part of the heart-stage embryo. Here, a file of cells is generated along the axis of the embryo, caused by impaired control over the planes of division of the hypophysis. As a result, the embryonic meristem fails to form correctly, although *monopteros* mutants can be induced to root experimentally. In contrast to *rootless* mutants, vascularization patterns in the cotyledons are often severely disrupted. Allelism studies between *rootless* and *monopteros* mutants are in progress. Since some hypocotyl and root development occurs in *rootless*, we do not consider it to be a pattern deletion mutant. Like *bashful* and *golftee*, the *rootless* mutant also co-segregates with a T-DNA. GUS fusion activity is detectable at very low levels in the vascular tissues.

hydra mutant. The fourth mutant we describe here, designated *hydra*, was also identified in a seedling screen and, like *emb30*, appears to show impairments of basic processes of cellular organization that affect several aspects of development. Seedlings homozygous for the mutation, which co-segregates with a T-DNA, are characterized by the presence of (typically five) green primordia surrounding the shoot apex, and a much reduced hypocotyl and root (Figure 4.6e). An investigation of the vascularization of the green leaf-like structures revealed it to be similar to that characteristic of cotyledons, although abnormal in patterning, and we conclude that the mutant produces multiple (almost invariably five) distorted cotyledons. The mutant is usually lethal to seedlings, but may occasionally produce flowers. During embryogenesis, heart-stage embryos lack the bilateral symmetry characteristic of the wild type, but five cotyledonary primordia of irregular size, shape and position are present. No embryonic root is evident in the mature embryo (Figure 4.7a). Anatomical studies have revealed that cell shape is abnormal from early in embryogenesis and, like *emb30*, homozygous mutant siblings show a diversity of phenotypes. Mutant seedlings also show defects in the organization of the vascular tissues, in the hypocotyl and root as well as in the cotyledons, with multiple vascular strands and poor connections between cells within the strands. Interestingly, individuals that are heterozygous for the *hydra* mutation also exhibit an abnormal phenotype, with a double vasculature in part of the root and no elongation zone. The mutation would therefore appear to be semi-dominant.

One mutant that bears certain similarities to *hydra* is *fass* (Torres-Ruiz and Jürgens, 1994), although complementation studies have revealed that the genes are not allelic (unpublished data). Alleles of the *fass* mutant, like *hydra*, are characterized by embryos composed of cells of a variety of sizes and shapes, and some alleles have supernumary cotyledons (three, four or five). Unlike the situation in the *hydra* mutant, the *fass* cotyledons are arranged in a characteristic II-shaped pattern and, while there is variability between siblings of particular *fass* alleles, reflected by inter-sibling variation in cotyledon number, our single allele of *hydra* shows almost no variation in cotyledon number, but varies in other aspects of shape, such as the site of cotyledon origin on the flanks of the embryonic axis, or the shape of the cotyledons. *fass* alleles identified to date are not lethal to seedlings, and the mutation is recessive.

4.5 Use of GUS fusion markers to analyse cell fate in mutants

One application of the cell-type-specific GUS fusion lines generated by promoter trapping is to use them to dissect processes of cell fate and cytodifferentiation in mutants which display defects in shape or pattern formation. We can illustrate this approach by reference to our studies on the establishment of apical-basal polarity in the *hydra* mutant.

One marker line we have used is AtEM101, which exhibits GUS fusion activity in the basal region of the embronic axis and, in the seedling, in the root meristem. This expression pattern therefore represents a molecular and visual marker of incipient or mature root meristem cells. This line has been crossed into *hydra* heterozygotes, and the expression of the marker was then investigated in F_2 *hydra* homozygous progeny. Analysis of embryos revealed that, even though the *hydra* cotyledonary stage embryo has no recognizable embryonic root, and is essentially globular in overall morphology, the GUS marker is expressed in the basal region. In older seedlings the marker is also expressed in cells that, in terms of their position, are expected to be root meristem cells, even though the organization of the root form is aberrant (Figure 4.7). These experiments clearly indicate that, even in the absence of correct shape, cells in the basal region of the *hydra* embryo are able to recognize their relative position and exhibit at least some features of the root differentiation pathway that are characteristic of the wild-type embryo and seedling.

4.6 Discussion and summary

One might intuitively expect that a precise co-ordination of the rate and plane of cell division and cell expansion during embryogenesis would be an absolute prerequisite for establishing correct pattern formation, morphogenesis and cytodifferentiation. Certainly, development of the wild-type embryo of *Arabidopsis* is characterized by a series of highly predictable cell divisions, indicative of a very tightly regulated series of cellular events. However, this is in contrast to other species such as maize, in which the generation of regular files of cells, in precise positions, is not a feature of embryogenesis (discussed by Lindsey and Topping, 1993). The following questions arise. What is the relationship between cell division/expansion, pattern formation, morpho-genesis and cytodifferentiation in *Arabidopsis* embryo development? Are these processes regulated as independent components that coincide passively during embryogenesis, or are they completely dependent on one another for the production of an integrated pathway of development?

The mutant studies summarized above allow us to draw a number of conclusions about these relationships. Perhaps the principal point is that the development of the embryo, and subsequently of the seedling, can be considered to be *modular*. That is, evidence has emerged that elements of development such as pattern formation, morphogenesis and cytodifferentiation can be dissected genetically, and must therefore be regulated, to some extent at least, independently. For example, the *fusca* mutants show defects in several aspects of differentiation, for example in the biosynthesis of anthocyanins, chlorophyll and various storage products and, in some alleles, in epidermal cell differentiation, but embryo morphology is unaffected. Similarly, the phenotypically 'immature' embryos and the suspensors of the *sus* and *raspberry* mutants can exhibit expression of genes that are normally characteristic of mature and morphologically fully developed embryos. These observations indicate that the differentiation of specific cell types is not absolutely linked to the generation of correct form, and 'chronological' markers of development, as represented by the normally late-appearing storage products, can be activated in the absence of correct morphology, and even in the wrong tissue. Certainly, genes have been identified that are determinants of the differentiation of specific cell types, functioning in a cell-autonomous way (e.g. the *GLABROUS1* gene which regulates trichome differentiation in *Arabidopsis*, see Herman and Marks, 1989; the *MIXTA* gene that is essential for corolla-lobe cell shape in *Antirrhinum*, see Noda *et al.*, 1994). In both of these cases, the genes have been cloned following insertional mutagenesis, and have been found to encode Myb-like transcription factors. Therefore, the fate of individual cells can clearly be altered autonomously in the absence of changes to organ morphology.

A different way of looking at the problem is to ask what the consequences are for correct cell differentiation if pattern formation or morphogenesis are disrupted. If we approach the problem from this angle, then we can identify genes that may be required not only for the establishment of correct shape, but also for histogenesis (tissue patterning) and cell differentiation, and we can further probe the relationship between these processes. Mutants such as *emb30/gnom*, *hydra*, *fass*, *rootless* and *monopteros* are clearly defective in shape, but they also all show evidence of abnormalities in tissue differentiation, particularly evident as the disruption of vascular tissue organization. In other words, the respective gene products are, in the wild-type plant, essential not only for morphogenesis but also for histogenesis, either directly or indirectly. All the mutants mentioned do, for example, produce at least some of the cell types of the vascular tissues, but the relative positioning of those cells, a basic component of tissue formation, is defective. This suggests that, while the differentiation of individual cell types may proceed independently of morphogenesis, the patterning of cells to form recognizable tissues may be dependent on cell–cell signalling events that are set up during morphogenesis.

A further question concerns the nature of the relationship between pattern formation and morphogenesis. This arises from observations of mutants such as *fass* and *hydra*, which retain elements of pattern (e.g. apical-basal polarity) in the absence of correct cellular organization. The establishment of polarity can therefore be separated genetically from the generation of correct shape (morphogenesis). In *hydra*, *fass* and *emb30/gnom* mutants, for example, homozygous siblings of single mutant alleles can show enormous variability in form, which may be due to chance events in the partitioning of cells during early divisions. However, elements of apical-basal polarity remain, except apparently (and as determined solely by anatomical criteria) in the more extreme *emb30* phenotypes, which are ball-shaped. These observations bring to mind ideas proposed by Haber (1962), who reported that irradiated wheat seedlings were able to establish organ polarity in the absence of any subsequent cell divisions whatsoever. Similarly, Foard *et al.* (1965) found that lateral roots can initiate in the absence of divisions of the pericycle. It would therefore appear that the relationship between correct cell division profiles and morphogenesis is not absolute.

If continuously organized patterns of cell division can be considered to be redundant with regard to pattern formation, as has been proposed by Kaplan and Hagemann (1991), then how can we interpret the mutant phenotypes described here? In the *hydra* mutant, for example, there is a lack of correlation between correct cellular organization and the establishment of apical-basal polarity, and these processes can be considered independently of one another. However, the shape of the embryo is clearly aberrant. Can we therefore consider the *hydra* shape to be abnormal, as a consequence of the observed lack of establishment of correct cell files? Our current working hypothesis is that the HYDRA protein is not required for cell division *per se*, but is involved in the control of correct cell shape, possibly because of a defect in the determination of the plane of cell expansion (e.g. through abnormal cell wall construction?), which in turn may influence the ability of the embryo to establish properly the relative positions of the fundamental tissue layers. Cell expansion, which is the primary contributor to plant growth, should therefore be considered a key determinant of plant shape and, in the *hydra* mutant, abnormal embryo shape may be the consequence of a loss of control of correct orientation of cell expansion. These questions are being addressed further in our laboratory.

There is undoubtedly a crucial role for correct cell division planes during *Arabidopsis* embryogenesis, and other aspects of *Arabidopsis* development, in contrast to the generalized views of Haber (1962) and Kaplan and Hagemann (1991). For example, the root-defective *monopteros* phenotype is correlated with abnormal planes of division of the hypophysis (Berleth and Jürgens, 1993), and in the transition to flowering, plants dramatically alter the patterns of cell division in the shoot apex prior to floral organ initiation, recruiting cells from the central region of the apical dome (Steeves and Sussex, 1989). While in many species correct cellular organization is characteristic of meristems, from which organ growth proceeds, it would appear not to be essential for at least some aspects of pattern formation. There is increasingly convincing evidence that the patterning of cell organization during embryogenesis provides a basic blueprint which, through cell expansion, is propagated during post-embryonic growth to generate characteristic cell lineages (seen exquisitely, for example, in the root, see Dolan *et al.*, 1993; also seen in the maize leaf, see Smith and Hake, 1992). Furthermore, corresponding partitioning of the embryonic symplast, in concert with the local activity (or indeed, activation) of cell-determining genes, may be crucial for creating the division of labour that is embryonic histogenesis. The division of cells partitions the cytoplasm and any regulatory molecules contained therein, such that the products of asymmetrical divisions may have quite different fates and gene expression patterns. This is seen not only in the *Arabidopsis* embryo, but has also been demonstrated directly in the angiosperm pollen grain (Eady *et al.*, 1995). One must also incorporate in this view other, supra-cellular mechanisms whereby organ pattern and shape are maintained, since cell lineage can be disrupted in maize mutants with no effect on organ shape (Dawe and Freeling, 1991).

These ideas are in stark contrast to the concepts that pervade animal developmental biology, where a primary function of cell division is considered to be the creation of building blocks. Further analyses of mutants will undoubtedly provide a greater insight into this fascinating problem of the co-ordination of plant development.

In summary, we can state that mutagenesis has proved to be an extremely informative experimental strategy for investigation of complex interactions in plant development. Even in the absence of cloned genes it is possible to gain significant insights into the hierarchies of organization of components of development. However, further information, particularly with regard to molecular mechanisms of pattern formation and cell differentiation, will require a knowledge of the nature of specific gene products, and of signalling molecules that are required to activate genes at the top of regulatory cascades. Significant progress is being made in specific experimental systems such as maize leaf development and *Arabidopsis* and *Antirrhinum* flower development. It should be apparent that the technique of insertional mutagenesis has advantages both in generating phenotypic mutations of interest and in facilitating the cloning of the mutant gene, by acting as a molecular tag. A modification of insertional mutagenesis is promoter trapping, which has a threefold advantage. First, it identifies and facilitates the cloning of genes that, when mutagenized, do not give rise to a phenotypically obvious developmental lesion. Secondly, it provides information about the spatial and temporal expression of a tagged gene throughout the life cycle of mutants that are lethal to embryos or seedlings in the homozygous state (the dominant *gus* reporter gene can be assayed in viable heterozygotes). Thirdly, it generates cell-type-specific markers that are valuable in analysing cell fate in particular mutant backgrounds. We expect that the combined use of promoter trapping and insertional mutagenesis, mediated by either T-DNA or transposon vectors, will become more widely used in the future for dissection of not only embryonic but also other developmental pathways in plants.

Acknowledgements

We wish to thank Dr Diane Shevell and Prof. Nam-Hai Chua for supplying *emb30* seed for allelism studies, Dr Brian Schwartz and Prof. David Meinke for *sus1-3* seed, and Dr Ben Scheres for *fass* seed and for many interesting discussions. We are also grateful to the Biotechnology and Biological Sciences Research Council and the Commission of the European Communities for financial support of our research on *Arabidopsis* development.

References

Aarts, M.G.M., Dirkse, W.G., Stiekema, W.J. and Pereira, A. (1993) Transposon tagging of a male sterility gene in *Arabidopsis*. *Nature* **363**, 715–717.

Bancroft, I., Jones, J.D.G. and Dean, C. (1993) Heterologous transposon tagging of the DRL1 locus in *Arabidopsis*. *Plant Cell* **5**, 631–638.

Bäumlein, H., Miséra, S., Luerßen, H., Kölle, K., Horstmann, C., Wobus, U. and Müller, A. (1994) The *FUS3* gene of *Arabidopsis thaliana* is a regulator of gene expression during late embryogenesis. *Plant J.* **6**, 379–387.

Becraft, P.W. and Freeling, M. (1991) Sectors of *liguleless-1* tissue interrupt an inductive signal during maize leaf development. *Plant Cell* **3**, 801–807.

Berleth, T. and Jürgens, G. (1993) The role of the *monopteros* gene in organising the basal body region of the *Arabidopsis* embryo. *Development* **118**, 575–587.

Bewley, J.D. and Black, M. (1994) *Seeds: Physiology of Development and Germination*, 2nd edn. Plenum Press, New York, London.

Castle, L.A. and Meinke, D.W. (1994) A *FUSCA* gene of *Arabidopsis* encodes a novel protein essential for plant development. *Plant Cell* **6**, 25–41.

Chaudhury, A.M., Letham, S., Craig, S. and Dennis, E.S. (1993) *amp1* — a mutant with high cytokinin levels and altered embryonic pattern, faster vegetative growth, constitutive photomorphogenesis and precocious flowering. *Plant J.* **4**, 907–916.

Clarke, M.C., Wei, W. and Lindsey, K. (1992) High frequency transformation of *Arabidopsis thaliana* by *Agrobacterium tumefaciens*. *Plant Mol. Biol. Rep.* **10**, 178–189.

Coen, E.S. and Meyerowitz, E.M. (1991) The war of the whorls: genetic interactions controlling flower development. *Nature* **353**, 31–37.

Dawe, R.K. and Freeling, M. (1991) Cell lineage and its consequences in higher plants. *Plant J.* **1**, 3–8.

Deng, X.-W., Matsui, M., Wei, N., Wagner, D., Chu, A.M., Feldmann, K.A. and Quail, P.A. (1992) *COP1*, an *Arabidopsis* regulatory gene, encodes a protein with both a zinc-binding motif and a Gb homologous domain. *Cell* **71**, 791–801.

Dolan, L., Janmaat, K., Willemsen, V., Linstead, P., Poethig, S., Roberts, K. and Scheres, B. (1993) Cellular organisation of the *Arabidopsis thaliana* root. *Development* **119**, 71–84.

Dolan, L., Duckett, C.M., Grierson, C., Linstead, P., Schneider, K., Lawson, E., Dean, C., Poethig, S. and Roberts, K. (1994) Clonal relationships and cell patterning in the root epidermis of *Arabidopsis*. *Development* **120**, 2465–2474.

Dure, III, L., Crouch, M., Harada, J., Ho, T.-H. D., Mundy, J., Quatrano, R., Thomas, T. and Sung, R. (1989) Common amino acid sequence domains among the *lea* proteins of higher plants. *Plant Mol. Biol.* **12**, 475–486.

Eady, C., Lindsey, K. and Twell, D. (1995) The significance of microspore division asymmetry for vegetative cell-specific transcription and generative cell differentiation. *Plant Cell* **7**, 65–74.

Feldmann, K.A. (1991) T-DNA insertion mutagenesis in *Arabidopsis*: mutational spectrum. *Plant J.* **1**, 71–82.

Foard, D.E., Haber, A.H. and Fishman, T.N. (1965) Initiation of lateral root primordia without completion of mitosis and without cytokinesis in uniseriate pericycle. *Am. J. Bot.* **52**, 580–590.

Fobert, P.R., Labbé, H., Cosmopoulos, J., Gottlob-McHugh, S., Ouellet, T., Hattori, J., Sunohara, G., Iyer, V.N. and Miki, B.L. (1994) T-DNA tagging of a seed coat-specific cryptic promoter in tobacco. *Plant J.* **6**, 567–577.

Franzmann, L.H., Yoon, E.S. and Meinke, D.W. (1995) Saturating the genetic map of *Arabidopsis thaliana* with embryonic mutations. *Plant J.* **7**, 341–350.

Furner, I.J. and Pumfrey, J.E. (1992) Cell fate in the shoot apical meristem of *Arabidopsis thaliana*. *Development* **115**, 755–764.

Giraudat, J., Hauge, B., Valon, C., Smalle, J., Parcy, F. and Goodman, H. (1992) Isolation of the *Arabidopsis ABI3* gene by positional cloning. *Plant Cell* **4**, 1251–1261.

Goldberg, R.B., Barker, S.J. and Perez-Grau, L. (1989) Regulation of gene expression during plant embryogenesis. *Cell* **56**, 149–160.

Haber, A. H. (1962) Nonessentiality of concurrent cell divisions for degree of polarization of leaf growth. I. Studies with radiation-induced mitotic inhibition. *Am. J. Bot.* **49**, 583–589.

Herman, P.L. and Marks, M.D. (1989) Trichome development in *Arabidopsis thaliana*. II. Isolation and complementation of the *GLABROUS1* gene. *Plant Cell* **1**, 1051–1055.

Hoch, M. and Jäckle, H. (1993) Transcriptional regulation and spatial patterning in *Drosophila*. *Curr. Opin. Genet. Dev.* **3**, 566–573.

Irish, V.F. and Sussex, I.M. (1992) A fate map of the *Arabidopsis* shoot apical meristem. *Development* **115**, 745–753.

Jefferson, R.A., Kavanagh, T.A. and Bevan, M.W. (1987) GUS fusions: β-glucuronidase as a sensitive and versatile gene fusion marker in higher plants. *EMBO J.* **6**, 3901–3907.

Kaplan, D.R. and Hagemann, W. (1991) The relationship between cell and organism in vascular plants. *BioScience* **41**, 693–703.

Keith, K., Kraml, M., Dengler, N.G. and McCourt, P. (1994) *fusca3*: a heterochronic mutation affecting late embryo development in *Arabidopsis*. *Plant Cell* **6**, 589–600.

Kertbundit, S., De Greve, H., DeBoeck, F., van Montagu, M. and Hernalsteens, J.-P. (1991) *In vivo* random β-glucuronidase gene fusions in *Arabidopsis thaliana*. *Proc. Natl Acad. Sci. USA* **88**, 5212–5216.

Koncz, C., Martini, N., Mayerhofer, R., Koncz-Kalman, Z., Körber, H., Rédei, G.P. and Schell, J. (1989) High frequency T-DNA-mediated gene tagging in plants. *Proc. Natl Acad. Sci. USA* **86**, 8467–8471.

Kornberg, T.B. and Tabata, T. (1993) Segmentation of the *Drosophila* embryo. *Curr. Opin. Genet. Dev.* **3**, 585–593.

Larkin, J.C., Oppenheimer, D.G., Pollock, S. and Marks, M.D. (1993) *Arabidopsis GLABROUS1* gene requires downstream sequences for function. *Plant Cell* **5**, 1739–1748.

Leyser, H.M.O., Lincoln, C.A., Timpte, C., Lammer, D., Turner, J. and Estelle, M. (1993) *Arabidopsis* auxin-resistance gene AXR1 encodes a protein related to ubiquitin-activating enzyme E1. *Nature* **364**, 161–164.

Lindsey, K. and Topping, J.F. (1993) Embryogenesis — a question of pattern. *J. Exp. Bot.* **44**, 359–374.

Lindsey, K., Wei, W., Clarke, M.C., McArdle, M.F., Rooke, L.M. and Topping, J.F. (1993) Tagging genomic sequences that direct transgene expression by activation of a promoter trap in plants. *Transgenic Res.* **2**, 33–47.

Liu, C.-M., Xu, Z.-H. and Chua, N.-H. (1993) Auxin polar transport is essential for the establishment of bilateral symmetry during early plant embryogenesis. *Plant Cell* **5**, 621–630.

Mayer, U., Torres-Ruiz, R.A., Berleth, T., Miséra, S. and Jürgens, G. (1991) Mutations affecting body organization in the *Arabidopsis* embryo. *Nature* **353**, 402–407.

Mayer, U., Büttner, G. and Jürgens, G. (1993) Apical-basal pattern formation in the *Arabidopsis* embryo: studies on the role of the *gnom* gene. *Development* **117**, 149–162.

Meinke, D. W. (1986) Embryo-lethal mutants and the study of plant embryo development. *Oxf. Surv. Plant Molec. Cell Biol.* **3**, 122–165.

Meinke, D.W. (1992) A homoeotic mutant of *Arabidopsis thaliana* with leafy cotyledons. *Science* **258**, 1647–1650.

Meinke, D.W. and Sussex, I.M. (1979) Embryo-lethal mutants of *Arabidopsis thaliana*: a model system for genetic analysis of plant embryo development. *Dev. Biol.* **72**, 50–61.

Meinke, D., Patton, D., Shellhammer, J., Reynolds-Duffer, A., Franzmann, L., Schneider, T. and Robinson, K. (1989) Development and molecular genetics of embryogenesis *Arabidopsis thaliana*. In (ed. R. Goldberg). *The Molecular Basis of Plant Development*. Alan R. Liss, New York, pp. 121–132.

Miséra, S., Müller, A.J., Weilandheidecker, U. and Jürgens, G. (1994) The *fusca* genes of *Arabidopsis*: negative regulators of light responses. *Mol. Gen. Genet.* **244**, 242–252.

Mullins, M.C. and Nüsslein-Volhard, C. (1993) Mutational approaches to study embryonic pattern formation in the zebrafish. *Curr. Opin. Genet. Dev.* **3**, 648–654.

Noda, K., Glover, B.J., Linstead, P. and Martin, C. (1994) Flower colour intensity depends on specialized cell shape controlled by a Myb-related transcription factor. *Nature* **369**, 661–664.

Okamuro, J.K., den Boer, B.G.W. and Jofuku, K.D. (1993) Regulation of *Arabidopsis* flower development. *Plant Cell* **5**, 1183–1193.

Priess, J.R. (1994) Establishment of initial asymmetry in early *Caenorhabditis elegans* embryos. *Curr. Opin. Genet. Dev.* **4**, 563–568.

Reiser, L. and Fischer, R.L. (1993) The ovule and the embryo sac. *Plant Cell* **5**, 1291–1301.

Scheres, B., Wolkenfelt, H., Willemsen, V., Terlouw, M., Lawson, E., Dean, C. and Weisbeek, P. (1994) Embryonic origin of the *Arabidopsis* primary root and root meristem initials. Development **120**, 2475–2487.

Schwartz, B.W., Yeung, E.C. and Meinke, D.W. (1994) Disruption of morphogenesis and transformation of the suspensor in abnormal *sus*pensor mutants of *Arabidopsis*. *Development* **120**, 3235–3245.

Shevell, D.E., Leu, W.-M., Gillmour, C.S., Xia, G., Feldmann, K.A. and Chua, N.-H. (1994) *EMB30* is essential for normal cell division, cell expansion, and cell adhesion in *Arabidopsis* and encodes a protein that has similarity to Sec7. *Cell* **77**, 1051–1062.

Smith, L.G. and Hake, S. (1992) The initiation and determination of leaves. *Plant Cell* **4**, 1017–1027.

Steeves, T.A. and Sussex, I.M. (1989) *Patterns in Plant Development*, 2nd edition. Cambridge University Press, Cambridge.

Thomas, T.L. (1993) Gene expression during plant embryogenesis and germination: an overview. *Plant Cell* **5**, 1401–1410.

Topping, J.F. and Lindsey, K. (1995) Insertional mutagenesis and promoter trapping in plants for the isolation of genes and the study of development. *Transgenic Res.* **4**, 291–305.

Topping, J.F., Wei, W. and Lindsey, K. (1991) Functional tagging of regulatory elements in the plant genome. *Development* **112**, 1009–1019.

Topping, J.F., Agyeman, F., Henricot, B. and Lindsey, K. (1994) Identification of molecular markers of embryogenesis in *Arabidopsis thaliana* by promoter trapping. *Plant J.* **5**, 895–903.

Torres-Ruiz, R.A. and Jürgens, G. (1994) Mutations in the *FASS* gene uncouple pattern formation and morphogenesis in *Arabidopsis* development. *Development* **120**, 2967–2978.

Tsay, Y.-F., Frank, M.J., Page, T., Dean, C. and Crawford, N.M. (1993) Identification of a mobile endogenous transposon in *Arabidopsis thaliana. Science* **260**, 342–344.

Walden, R., Hayashi, H. and Schell, J. (1991) T-DNA as a gene tag. *Plant J.* **1**, 281–288.

Wei, N. and Deng, X.-W. (1993) Characterisation and molecular cloning of COP9, a genetic locus involved in light-regulated development and gene expression in *Arabidopsis. In: Proceedings of the 5th International Conference on* Arabidopsis *Research* (eds R. Hangarter, R. Scholl, K. Davis and K. Feldmann). Ohio State University, Columbus, Ohio, p. 14.

Wei, N., Kwok, S.F., von Arnim, A.G., Lee, A., McNellis, T.W., Piekos, B. and Deng, X.-W. (1994a) *Arabidopsis COP8, COP10* and *COP11* genes are involved in repression of photomorphic development in darkness. *Plant Cell* **6**, 629–643.

Wei, W., McArdle, H. and Lindsey, K. (1994b) ΔgusBin19 T-DNA. *Plant Mol. Biol.* **26**, 1021.

West, M.A.L. and Harada, J.J. (1993) Embryogenesis in higher plants: an overview. *Plant Cell* **5**, 1361–1369.

West, M., Yee, K.M., Danao, J., Zimmerman, J.L., Fischer, R.L., Goldberg, R.B. and Harada, J.J. (1994) *LEAFY COTYLEDON1* is an essential regulator of late embryogenesis and cotyledon identity in *Arabidopsis. Plant Cell* **6**, 1731–1745.

Whitelam, G.C., Johnson, E., Peng, J., Carol, P., Anderson, M., Cowl, J.S. and Harberd, N.P. (1993) Phytochrome A null mutants of *Arabidopsis* display a wild-type phenotype in white light. *Plant Cell* **5**, 757–768.

Yadegari, R., de Paiva, G.R., Laux, T., Koltunow, A.M., Apuya, N., Zimmerman, J.L., Fischer, R.L., Harada, J.J. and Goldberg, R.B. (1994) Cell differentiation and morphogenesis are uncoupled in *Arabidopsis raspberry* embryos. *Plant Cell* **6**, 1713–1729.

Yeung, E.C. and Meinke, D.W. (1993) Embryogenesis in angiosperms: development of the suspensor. *Plant Cell* **5**, 1371–1381.

5

Protein kinase genes expressed during plant reproductive development

J. Tregear, V. Decroocq-Ferrant, S. Jouannic and M. Kreis

5.1 Introduction

In both yeast and the fruit fly *Drosophila*, much progress has been made over the last few years in the search for genes involved in development. The temporal and spatial expression patterns of various developmental genes have been determined, and the mechanisms by which many of these genes influence the regulation of other genes have been elucidated. As a result, cascades of regulatory genes have been identified in which the product of each gene regulates the activity of another gene at a defined site and at a discrete time in development. The influence of these findings on our understanding of developmental biology is profound, because they reveal mechanisms of gene interaction and regulation which might be conserved in a wide range of organisms.

The use by diverse eukaryotic organisms of evolutionarily related genes in the control of development is perhaps best illustrated by protein kinase genes, which may be conserved in the form of signal transduction modules representing entire regulatory cascades. The evolutionary conservation of protein kinases in modular form is probably best illustrated by the mitogen-activated protein (MAP) kinase cascade-type pathway found in each of the three main eukaryotic groups (see Figure 5.1 and Section 5.2).

Many of the first protein kinase genes which were found to be important in the regulation of development were identified in yeast, *Drosophila* and other organisms amenable to genetic analysis. In contrast, relatively few plant genes of this kind have been isolated and analysed in detail. However, the remarkable structural and functional conservation of an appreciable number of protein kinases and signalling pathways, such as the MAP kinase pathway, offers a way to isolate plant homologues of developmentally important protein kinase genes using the polymerase chain reaction (PCR). Molecular genetic and cell biology approaches then provide the means to study in depth the function of these novel plant genes.

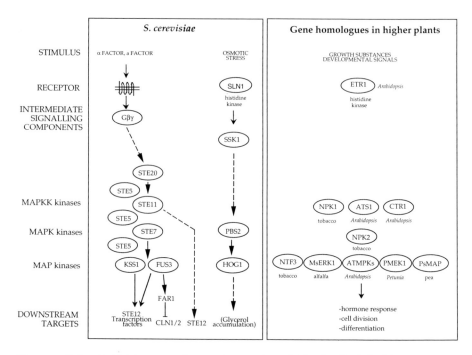

Figure 5.1. *MAP kinase pathways in yeast and plants. Protein kinases involved in these well studied pathways of yeast (Hughes, Herskowitz, 1995) are shown together with homologous of these kinases which have been identified in higher plants (see references in Section 5.2).*

Because of our interest in gametogenesis and embryogenesis in higher plants, we have been searching for plant homologues of protein kinases which are known to play an important role during reproduction and embryogenesis in other organisms such as *Xenopus laevis* and *Drosophila melanogaster*. We are particularly interested in plant homologues of genes that specify elements of MAP kinase pathways and genes homologous to the segment polarity gene *shaggy* of *Drosophila*. In order to isolate plant homologues of the genes mentioned above, a PCR-based homology cloning approach (Bianchi *et al.*, 1993, 1994; Kreis *et al.*, 1994) was used. Mixtures of oligonucleotides specifying highly conserved subdomains of protein serine/threonine kinases (Hanks *et al.*, 1988; Hanks, 1991) were used for the amplification of cDNA synthesized from flower bud, ovule and embryo sac mRNA. The amplification products obtained were found to contain open reading frames possessing the characteristic hallmarks of protein serine threonine kinases. PCR products were subsequently used as probes to screen cDNA libraries for the purpose of obtaining full length clones.

5.2 MAP kinase-related genes

Mitogen-activated protein (MAP) kinases (also referred to as external-signal-regulated protein kinases (ERKs)) constitute a subfamily of protein kinases which are highly conserved in eukaryotes ranging from yeast to mammals, *Drosophila* and plants (Blenis, 1993; Jonak *et al.*, 1994a; Neiman, 1993; Tregear *et al.*, 1995). The activation of MAP kinases occurs via phosphorylation on tyrosine and threonine residues of the conserved TEY motif (see Table 5.1), mediated by a dual specific MAP kinase kinase (MAPKK) (Ahn *et al.*, 1992). The latter kinase is activated by phosphorylation on serine and threonine residues by another protein kinase, MAP kinase kinase kinase (MAPKKK), for which two classes, MEKK and Raf, have been identified (Lange-Carter *et al.*, 1993). The MAPK/MAPKK/ MAPKKK module has been identified in yeast (see Errede and Levin, 1993), mammals (Boulton *et al.*, 1991; Brand and Perrimon, 1994) and, more recently, in plants (Banno *et al.*, 1993; Shibata *et al.*, 1995), and seems to be involved in diverse signal transduction pathways. These regulatory cascades, which have been shown in many cases to play an important role in the transduction of developmental signals, illustrate the great versatility of MAP kinase-mediated signalling pathways. For example, recent results obtained by Brand and Perrimon (1994) suggest that a MAP kinase kinase may function downstream of the EGF (epidermal growth factor) receptor and the Raf kinase to transduce a signal that specifies follicle-cell fate and the establishment of polarity in the developing *Drosophila* embryo. Similarly, in *Xenopus laevis* the product of the *c-mos* gene (a serine/threonine kinase which phosphorylates the Raf MAPKKK) was found to be germ cell specific, synthesized from stored mRNAs in response to progesterone, and functioning exclusively during gametogenesis and embryogenesis (Sagata *et al.*, 1989).

Plant MAP kinase genes have been identified in Alfalfa, *Arabidopsis*, pea and

Table 5.1. *Sequence comparison of MAP kinases (see Table 5.2) from different plant species between subdomains VI and VIII*[a]

	VI	VII	° °	VIII
MsERK1	**HRD**LKPSNLLLNANCDLKIC**DFG**LARVTSETD--FM**TEY**VVTRWYR**APE**			
D5	**HRD**LKP*N**************DFG***********--****TEY********* **APE**			
NTF3	**HRD**LKPGN**I************DFG*****TSSGKDQ-****TEY** ******* **APE**			
ATMPK1	**HRD**LKPGN**V********* **DFG*****ASNTKGQ-****TEY** ******* **APE**			
ATMPK2	**HRD**LKPGN**V********* **DFG*****TSNTKGQ-****TEY** ******* **APE**			
ATMPK3	**HRD**LKP*N *********** **DFG*****PT**N*--****TEY** ******* **APE**			
ATMPK4	**HRD**LKP*N *********LG **DFG***** TK****--****TEY** ******* **APE**			
ATMPK5	**HRD**LKP*N ****S******T**DFG** ***T****E--** **TEY** ******* **APE**			
ATMPK6	**HRD**LKP*N *********** **DFG** ***V***S*--****TEY** ******* **APE**			
ATMPK7	**HRD**LKPGN**V***********DFG*****TSQGNEQ-****TEY** ******* **APE**			
PMEK1	**HRD**LKPGN**I***********DFG*****TSSGKDQ-****TEY** ******* **APE**			
ERK2	**HRD**LKPSNLLLNTTCDLKIC**DFG**LARVADPDHTGFL**TEY**VATRWYR**APE**			

[a]Rat ERK2 (Boulton *et al.*, 1991) is added as a reference for non-plant organisms. The invariant residues typical of MAP kinases, within the motif TEY, are designated by (°) and shown in bold. Identical residues are denoted by an asterisk.

Table 5.2. *Similarities between plant members of the MAP kinase family and the rat ERK2 and alfalfa MsERK1 MAP kinases*

MAP kinase	Plant	ERK2	MsERK1	References
MsERK1	Alfalfa	51.6	*	Duerr *et al.*, 1993
MsK7	Alfalfa	51.6	100	Jonak *et al.*, 1993[a]
MsK11	Alfalfa	?	?	Jonak *et al.*, 1994
MsK14	Alfalfa	?	?	Jonak *et al.*, 1994
MsK17	Alfalfa	?	?	Jonak *et al.*, 1994
D5	Pea	51	94.4	Stafstrom *et al.*, 1993
NTF3	Tobacco	50	54	Wilson *et al.*, 1993
ATMPK1	*Arabidopsis*	49.7	56.1	Mizoguchi *et al.*, 1994
ATMPK2	*Arabidopsis*	50	55.2	Mizoguchi *et al.*, 1994
ATMPK3	*Arabidopsis*	49.7	76.9	Mizoguchi *et al.*, 1993
ATMPK4	*Arabidopsis*	51.2	72.2	Mizoguchi *et al.*, 1993
ATMPK5	*Arabidopsis*	50.3	67.5	Mizoguchi *et al.*, 1993
ATMPK6	*Arabidopsis*	51.3	88.4	Mizoguchi *et al.*, 1993
ATMPK7	*Arabidopsis*	46.7	53.9	Mizoguchi *et al.*, 1993
ATMPK8	*Arabidopsis*	?	?	Mizoguchi *et al.*, 1994b
ATMPK9	*Arabidopsis*	?	?	Mizoguchi *et al.*, 1994b
PMEK1	*Petunia*	44	50	Decroocq-Ferrant *et al.*, 1995b

[a]MsK7 and MsERK1 were isolated independently but found to be identical. Question marks denote the absence of a published sequence. Rat ERK2 (Boulton *et al.*, 1991) is added as a reference for non-plant organisms.

tobacco (see Table 5.2). *NPK2*, a MAPKK homologous gene (Shibata *et al.*, 1995) and *NPK1*, a gene encoding a putative MAPKKK (Banno *et al.*, 1993), have been identified in tobacco, suggesting that higher plants contain entire MAP kinase signal transduction modules similar to those of yeast and animals. Although the functions of the plant MAP kinase signal transduction pathways are not known, preliminary evidence indicates that they might play a role in plant hormone signalling during auxin-induced cell division (Mizoguchi *et al.*, 1994a) and in the ethylene response (Kieber *et al.*, 1993).

As part of our study of the molecular control of ovule development and early embryogenesis in *Petunia hybrida*, we have investigated MAP kinase gene expression during late flower bud development. PCR was used to amplify a MAP kinase cDNA fragment from *P. hybrida* ovule and embryo sac cDNA (Decroocq-Ferrant *et al.*, 1995b; Ferrant *et al.*, 1994). A full-length clone, subsequently named *PMEK1*, was isolated by screening an ovule-specific cDNA library with the PCR fragment. The *PMEK1* cDNA codes for a protein of 44.4 kDa with an overall sequence identity of 44% to the product of the rat gene *erk2* (see Table 5.2). An alignment of the amino acid sequence of *PMEK1* with those of the MAP family of protein kinases from *Arabidopsis*, pea and alfalfa shows identities ranging from 44% to 96%. *PMEK1* is most closely related to *ATMPK1* and *NTF3* from *Arabidopsis* and tobacco, respectively. On the basis of these sequence identities, the PMEK1 kinase may be placed in the plant MAPK group I designated by Mizoguchi *et al.* (1993). In order to study the expression of *PMEK1* in reproductive organs of *P. hybrida*, poly(A) + RNA isolated from *P. hybrida* floral buds at different stages of flower bud development (for details, see Decroocq-Ferrant *et al.*, 1995b) was used in Northern and reverse transcript-PCR amplification RT-PCR analysis. In contrast with the results obtained by Wilson *et al.* (1993), who reported that the tobacco NTF3 gene is expressed in both female and male reproductive organs, our Northern experiments revealed the presence of *PMEK1* transcripts only in the female part of the *Petunia* flower. *In situ* hybridization was carried out on serial sections of *P. hybrida* floral buds from the ovule primordia stage through to embryo sac maturity in order to study the temporal and spatial expression pattern of the *PMEK1* gene. The results show that *PMEK1* transcripts are distributed throughout the ovule. The outer layer of the placenta showed significant hybridization, whilst the ovary and inner placenta exhibited only low levels of hybridization (see Decroocq-Ferrant *et al.*, 1995b).

In parallel with our studies on the *PMEK1* gene of *Petunia*, we have isolated cDNAs of another gene potentially involved in MAP kinase signalling in *Arabidopsis*. Using the PCR-based approach described above, we have isolated a cDNA, *ATS1*, which specifies a protein kinase which shares closest homology within its catalytic domain with the yeast MAPKKK homologue *BCK1* (Lee and Levin, 1992) (40.9% identity). The next closest relative is the tobacco *NPK1* gene, which is thus a member of the same plant gene family, although the divergence between these two kinases suggests that *ATS1* is not the *Arabidopsis* equivalent of *NPK1*. The *ATS1* cDNA was isolated from young shoot mRNA, and we are currently investigating its distribution in different organs of the plant, including reproductive tissues. We plan subsequently to express the kinase in a heterologous expression system in order to study its phosphorylation and raise antibodies for further studies at the protein level.

We are at the same time attempting to isolate clones for other *Arabidopsis* genes whose products might interact with those of the *ATS1* gene. We are particularly interested in isolating potential homologues of the yeast Ste20 protein kinase, which acts upstream of the MAPKK kinase Ste11 in the yeast mating pathway (Figure 5.1, Leberer *et al.*, 1992).

5.3 Plant genes related to the segment polarity gene *shaggy/zeste-white 3 (SGG/ZW3)* from *Drosophila*

The *Drosophila* sgg/zw3 protein kinase (Bourouis *et al.*, 1990; Siegfried *et al.*, 1990) and the rat GSK-3 protein kinase (Woodgett, 1990, 1991) belong to the GSK-3 subfamily of serine/threonine kinases (as defined by Hanks, 1991), which also includes MDS1 from yeast (Bianchi *et al.*, 1993; Puziss *et al.*, 1994). The latter kinase is an absolute requirement for the completion of meiosis. *sgg/zw3* is required for signalling mediated by the *wingless* gene-product (Perrimon, 1994), and plays a key role in the process of segmental patterning during early embryogenesis in *Drosophila*, as well as during bristle development. The process of segmental patterning occurs through cell–cell signalling to determine cell identity. *sgg/zw3* acts upstream of the homeotic gene engrailed (*en*), and specifies cell identity within each segmental unit of the *Drosophila* embryo. The protein kinase encoded by *sgg/zw3* is also required in a lateral signalling pathway downstream of the transmembrane protein Notch during bristle development. Hülskamp *et al.* (1994) and Larkin *et al.* (1994) have suggested that genes analogous to those invoved in *Drosophila* bristle development could operate during plant trichome development.

Recently, Ruel *et al.* (1993) showed that the rat *GSK-3* gene is capable of restoring *sgg/zw3* mutants to a wild-type phenotype. This conservation of function between the rat *GSK-3* and fruit fly *sgg/zw3* genes is an example of how well strategies of gene interaction may be conserved in a wide range of organisms. That this conservation extends across the three main groups of eukaryotes has been shown by the cloning of plant genes related to *sgg/zw3* and *GSK-3*.

Table 5.3 shows that the catalytic domain sequences of the protein kinases encoded by the *ASK*, *MsK* and *PSK* genes from *Arabidopsis*, alfalfa and *Petunia*, respectively, are about 70% identical to the corresponding kinase domains encoded by the *sgg/zw3* and *GSK-3* genes. Bianchi *et al.* (1994) showed that recombinant *ASK* proteins phosphorylate phosphatase inhibitor-2 and myelin basic protein *in vitro* on serine and threonine, respectively, in a similar manner to the GSK-3 kinase. Moreover, autophosphorylation was found to occur not only on threonine and serine residues but also, though less strongly, on

Table 5.3. *Comparison of plant shaggy like protein kinase sequences (Bianchi et al., 1994; Decroocq-Ferrant et al., 1995; Pay et al., 1993) with shaggy/zeste-white 3 from* Drosophila *(Siegfried et al., 1990; Bourouis et al., 1990), GSK-3 from rat (Woodgett, 1990) and MDSI from yeast (Puziss et al., 1994); the percentage identity between each pair of shaggy like proteins is shown*

		1	2	3	4	5	6	7
	1- PSK4	100						
	2- PSK6	80	100					
Arabidopsis	3- ASKα	90	81	100				
Alfalfa	4- MsK1	85	82	85	100			
	5- Shaggy	70	70	71	69	100		
	6- GSK-3	70	70	73	68	86	100	
Yeast	7- MDS1	54	55	60	57	60	60	100

phosphotyrosine. The *GSK-3* protein has in fact been shown to require phosphorylation on the tyrosine residue 216, which is conserved in all the *sgg/GSK-3* homologues characterized to date, in order to become active (Hughes *et al.*, 1993). These results suggest that the plant kinases probably have similar biochemical activities to their insect and mammalian counterparts.

In view of the fundamental role which the *GSK-3* homologues play during oogenesis (Bender *et al.*, 1993), early embryogenesis (He *et al.*, 1995) and meiosis (Puziss *et al.*, 1994), we decided to study the expression of related genes during sporogenesis, gametogenesis and embryogenesis in *Petunia*. Two full-length cDNA clones, *PSK4* and *PSK6*, encoding homologues of GSK-3-related protein kinases, were isolated from a *P. hybrida* ovule-specific library (Ferrant *et al.*, 1994; Table 5.3). Sequence comparisons suggest that PSK4 and PSK6 fall in to two subgroups of plant protein kinases related to the GSK-3 subfamily (see Decroocq-Ferrant *et al.*, 1995a).

Northern experiments by Decroocq-Ferrant *et al.* (1995a) showed that the *PSK4* gene is expressed during normal vegetative growth, whilst *PSK6* transcripts are barely detectable in leaf, root and stem tissues. Both genes are expressed in the ovule and in mature embryo sacs, but only *PSK6* transcripts accumulate to high levels in the anther. By means of *in situ* hybridization, it was demonstrated that *PSK4* transcripts are present in both the placenta and the ovule itself during megagametogenesis (Figure 5.2; Decroocq-Ferrant *et al.*, 1995a), their abundance gradually increasing during this period. No transcripts were detected in the male reproductive tissue. *In situ* hybridization using a PSK6 probe showed that, at the time of pollen meiosis, transcripts of this gene are localized almost exclusively in the cells surrounding the sporogenic tissue of the stamens. As expected, *PSK6* transcripts were found to accumulate transiently during anther and pollen differentiation, and were also detected in the tapetum before the completion of tetrad formation. A preliminary investigation of *PSK6* and *PSK4* gene expression in the young embryo revealed the presence of transcripts only at the globular stage.

5.4 Future prospects

Although protein kinases are known to be implicated in the regulation of a plethora of cellular processes and are essential regulatory elements in many signal transduction pathways, only a small number of these proteins have as yet been studied in detail in plants, and their roles are as yet poorly understood. Recently, various other components of signal transduction pathways, such as G proteins and phosphatidylinositol (4,5) bisphosphate (Ma, 1993), previously only known to be present in the other eukaryotic groups, have also been identified in plants. This suggests that protein kinases may represent only one part of a basic signalling machinery which is conserved to greater or lesser extents in all eukaryotes.

Protein phosphorylation in plants probably regulates, in addition to the universal pathways, various unique processes such as photosynthesis and photomorphogenesis (Allen, 1992). Protein phosphorylation is likely to be very important during various events in early and late flower development, in

FLOWER BUD STAGE B0: MALE SPOROGENESIS AND MEIOSIS AND OVULE PRIMORDIA

(a)   **(b)**

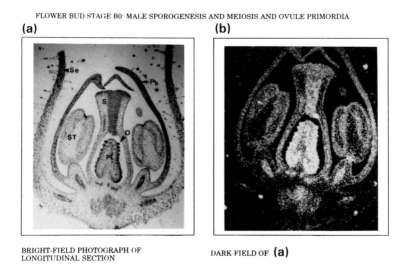

BRIGHT-FIELD PHOTOGRAPH OF
LONGITUDINAL SECTION   DARK-FIELD OF **(a)**

Figure 5.2. In situ *hybridization of* PSK4 *mRNA during flower organogenesis at stage B0 (Decroocq-Ferrant* et al., *1995a). Regions where RNA/RNA hybrids occur are visible as intense white dots visible above a background of grey dots. S, stigma; O, ovule; Pe, petal; PL, placenta; Se, sepal; ST, stamen.*

pollen/stigma recognition and in embryogenesis (Roe *et al.*, 1993; Walker and Zhang, 1990; Walker, 1993). By analogy with what is known of other systems, notably *Drosophila*, it is anticipated that the characterization of protein kinase genes expressed during gametogenesis and embryogenesis will provide insights into the nature of the signalling pathways implicated in cell–cell communication and in the determination of cell fate.

The picture which is emerging from studies on other organisms suggests that multiple branching of signal transduction chains and signalling element redundancy are likely to be commonplace. In the case of plant MAP kinase pathways, the identification of a minimum of seven different MAP kinase genes in *Arabidopsis* is not inconsistent with this prediction. Given the diverse roles of MAP kinase pathways in eukaryotes, it seems probable that both the receptor/ligand interactions feeding into these pathways and the target molecules acting downstream of them are much less conserved than are the MAP kinase modules themselves. Nevertheless, there is preliminary evidence that, like many of their counterparts in other organisms, plant MAP kinases may be involved in growth and differentiation. Mizoguchi *et al.* (1994a) have shown that recombinant *ATMPK1* and *ATMPK2* proteins (see Table 5.2) are phosphorylated and thereby activated in auxin-starved suspension cultures in response to auxin treatment. There is also evidence that differentiation in response to ethylene might involve a MAP kinase pathway, since the *Arabidopsis CTR1* gene involved in this response has been found to encode a homologue of the mammalian *raf* gene, which acts as a MAPKK kinase (Kieber *et al.*, 1993). Although it is as yet unclear whether there are any MAP kinase pathways which are specific to reproduction or embryogenesis, the general involvement of MAP kinase pathways in differentiation in other eukaryotes suggests that they are likely to play a role of some kind in these processes. It is interesting to note that the *ctr1* mutant phenotype includes the presence of elongated carpels which protrude from the unopened floral buds. This suggests an interaction between growth substance response and floral development pathways in which a MAP kinase cascade might be implicated. The *in situ* localization pattern of *PMEK1* transcripts in the developing floral tissues of *Petunia* is consistent with an important role for MAP kinase cascades in plant reproduction. The same conclusion is true for the *sgg/zw3*-related genes, and it will be interesting to determine whether the strikingly high degree of homology shared with their animal and yeast counterparts provides clues as to their exact function.

The pattern and timing of expression of the *Petunia PSK* genes during the development and maturation of the reproductive organs indicates that the *PSK* proteins may fulfil more than one function, depending on whether the cell in which they are expressed is of gametophytic or sporophytic origin. By analogy with processes observed in *Drosophila*, the PSK protein kinases could possibly be involved in communication between the sporophytic (i.e. tapetum and integuments) and the gametophytic cells (microspores and embryo sac) in the anther and in the ovule.

5.5 Conclusions

In recent years it has become apparent that protein kinases play a key role in signal transduction pathways, including those involved in reproductive development and embryogenesis. Hence, it is hoped that the characterization and investigation of the function of protein kinases expressed during the development of plant reproductive organs will provide information about the signalling pathways implicated in cell-type determination in these tissues and the possible role of the accumulated gene products during embryogenesis. Recent work in our laboratory has demonstrated the presence of *shaggy/GSK-3* and MAP kinase signalling element homologues in both *Petunia* and *Arabidopsis*. MAP kinase and *shaggy/GSK-3* expression have been studied in detail using *in-situ* hybridization, demonstrating that certain of the genes identified are selectively expressed in reproductive organs, their transcripts showing specific temporal and spatial accumulation patterns. These results suggest that the protein kinases encoded by the genes studied might indeed play a role during these stages of development.

Acknowledgements

We are grateful to Professors Jac van Went and Sacco de Vries for their valuable assistance, and we thank Roland Boyer (CNRS) for photographic work. James Tregear gratefully acknowledges EC support in the form of a Biotechnology Programme fellowship.

References

Ahn, N.G., Seger, R. and Krebs, E.G. (1992) The MAP kinase activator. *Curr. Opin. Cell Biol.* **4**, 992–999.

Allen,J.F. (1992) How does protein phosphorylation regulate photosynthesis? *Trends Biochem. Sci.* **17**, 12–17.

Banno, H., Hirano, K., Nakamura, T., Irie, K., Nomoto, S., Matsumoto, K. and Machida, Y. (1993) NPK1, a tobacco gene that encodes a protein with a domain homologous to yeast BCK1, STE11 and Byr2 protein kinases. *Mol. Cell. Biol.* **13**, 4745–4752.

Bender, L.B., Kooh, P.J. and Muskavitch, M.A.T. (1993) Complex function and expression of *Delta* during *Drosophila* oogenesis. *Genetics* **133**, 967–978.

Bianchi, M.W., Plyte, S.E., Kreis, M. and Woodgett,J.R. (1993) A *Saccharomyces cerevisiae* protein-serine kinase related to mammalian glycogen synthase kinase-3 and the *Drosophila melanogaster* gene shaggy product. *Gene* **134**, 51–356.

Bianchi, M.W., Guivarc'h, D., Thomas, M., Woodgett,J.R. and Kreis, M. (1994) *Arabidopsis* homologs of the *shaggy* and GSK-3 protein kinases: molecular cloning and functional expression in *E. coli. Mol. Gen. Genet.* **242**, 337–45.

Blenis, J. (1993) Signal transduction via MAP kinases: proceed at your own RSK. *Proc. Natl Acad. Sci. USA* **90**, 5889–5892.

Boulton, T.G., Nye, S.H., Robbins, D.J., Ip, N.Y., Radziejewska, E., Morgenbesser, S.D., DePinho, R.A., Panayotatos, N., Cobb, M.H. and Yancopoulos, G.D. (1991) ERKs, a family of protein-serine/threonine kinases that are activated and tyrosine phosphorylated in response to insulin and NGF. *Cell* **65**, 663–675.

Bourouis, M., Moore, P., Ruel, L., Grau, Y., Heitzler, P. and Simpson, P. (1990) An early embryonic product of the gene shaggy encodes a serine/threonine protein kinase related to the CDC28/cdc2 subfamily. *EMBO J.* **9**, 2877–2884.

Brand, A.H. and Perrimon, N. (1994) Raf acts downstream of the EGF receptor to determine dorsoventral polarity during *Drosophila* oogenesis. *Genes and Development* **8**, 629–639.

Decroocq-Ferrant, V., Van Went, J., de Vries, S., Bianchi, M.W. and Kreis, M. (1995a) *Petunia* hybrida homologues of shaggy/zeste-white 3 expressed in female and male reproductive organs. *Plant J.* **7**, 897–911.

Decroocq-Ferrant, V. Decroocq, S. Van Went, J. Schmidt, E. and Kreis, M. (1995b) A homologue of the MAP/ERK family of protein kinase genes is expressed in vegetative and in female reproductive organs of *Petunia hybrida. Plant Mol. Biol.* **27**, 339–350.

Duerr, B., Gawienowski, M., Ropp, T. and Jacobs, T. (1993) MsERK1, a mitogen-activated protein kinase from a flowering plant. *Plant Cell* **5**, 87–96.

Errede, B. and Levin, D.E. (1993) A conserved kinase cascade for MAP kinase activation in yeast. *Curr. Opin. Cell Biol.* **5**, 254–260.

Ferrant, V., Van Went, J. and Kreis, M. (1994) Ovule cDNA clones of *Petunia* hybrida encoding proteins homologous to MAP and shaggy/zeste-white 3 protein kinases. In: *Molecular and Cellular Aspects of Plant Reproduction* (eds R. Scott and A. Stead). Cambridge University Press, Cambridge, pp. 159–172.

Hanks, S.K. (1991) Eukaryotic protein kinases. *Curr. Opin. Struct. Biol.* **1**, 369–383.

Hanks, S.K., Quinn, A.M. and Hunter, T. (1988) The protein kinase family: conserved features and deduced phylogeny of the catalytic domains. *Science* **241**, 42–52.

He, X., Saint-Jeannet, J.P., Woodgett,J.R., Varmus, H.E. and Dawid, I.B. (1995) Glycogen synthase kinase-3 and dorsoventral patterning in *Xenopus* embryos. *Nature* **374**, 617–622.

Herskowitz, I. (1995) MAP kinase pathways in yeast: for mating and more. *Cell* **80**, 187–188.

Hughes, D.A. (1994) Histidine kinases hog the limelight. *Nature* **369**, 187–188.

Hughes, K., Nikolakaki, E., Plyte, S.E., Totty, N. and Woodgett, J.R. (1993) Modulation of the glycogen synthase kinase-3 family by tyrosine phosphorylation. *EMBO J.* **12**, 803–808.

Hülskamp, M., Miséra, S. and Jürgens, G. (1994) Genetic dissection of trichome cell development in *Arabidopsis*. *Cell* **76**, 555–566.

Jonak, C., Pay, A., Bögre, L., Hirt, H. and Heberle-Bors, E. (1993) The plant homologue of MAP kinase is expressed in a cell cycle-dependent and organ-specific manner. *Plant J.* **3**, 611–617.

Jonak, C., Heberle-Bors, E. and Hirt, H. (1994a) MAP kinases: universal multi-purpose signaling tools. *Plant Mol. Biol.* **24**, 407–416.

Jonak, C., Kiegerl, S., Bögre, E., Heberle-Bors, E. and Hirt, H. (1994b) A MAP kinase gene family in alfalfa. Poster abstract no. 877, Fourth International Congress of Plant Molecular Biology, Amsterdam, 1994.

Kieber, J., Rothenberg, M., Roman, G., Feldmann, K.A. and Ecker, J.R. (1993) CTR1, a negative regulator of the ethylene response pathway in *Arabidopsis*, encodes a member of the Raf family of protein kinases. *Cell* **72**, 427–441.

Kreis, M., Bianchi, M.W., Ferrant, V., Le Guen, L., Thomas, M., Halford, N.G., Barker, J.H.A., Hannappel, U., Vicente-Carbajosa, J. and Shewry, P.R. (1994) Plant genes encoding homologues of the SNF1 and shaggy protein kinases. In: *Molecular Biology: Molecular-Genetic Analysis of Plant Development and Metabolism* (eds G. Corruzi and P. Puigdomenech). NATO Asi series H81, Springer-Verlag, Berlin, pp. 453–467.

Lange-Carter, C.A., Pleiman, C.M., Gardner, A.M., Blumer, K.J. and Johnson, G.L. (1993) A divergence in the MAP kinase regulatory network defined by MEK kinase and Raf. *Science* **260**, 315–319.

Larkin, J.C., Oppenheimer, D.G., Lloyd, A.M., Paparozzi, E.T. and Marks, M.D. (1994) Roles of the glabrous and transparent testa glabra in *Arabidopsis* trichome development. *Plant Cell* **6**, 1065–1076.

Leberer, E., Dignard, D., Harcus, D., Thomas, D.Y. and Whiteway, M. (1992) The protein kinase homologue Ste20p is required to link the yeast pheromone response G-protein $\beta\gamma$ subunits to downstream signalling components. *EMBO J.* **11**, 4815–4824.

Lee, K.S. and Levin, D.E. (1992) Dominant mutation in a gene encoding a putative protein kinase (BCK1) bypass the requirement for a *Saccharomyces cerevisiae* protein kinase C homolog. *Mol. Cell. Biol.* **12**, 172–182.

Ma, H. (1993) Protein phosphorylation in plants: enzymes, substrates and regulators. *Trends Genet.* **9**, 228–230.

Mitchell, A.R. (1994) Control of meiotic gene expression in *Saccharomyces cerevisiae*. *Microbiol. Rev.* **58**, 56–70.

Mizoguchi, T. Hayashida, N., Yamaguchi-Shinozaki, K., Kamada, H. and Shinozaki, K. (1993) ATMPKs: a gene family of plant MAP kinases in *Arabidopsis thaliana*. *FEBS Lett.* **336**, 440–444.

Mizoguchi, T., Gotoh, Y., Nishida, E., Yamaguchi-Shinozaki, K., Hayashida, N., Iwasaki, T., Kamada, H. and Shinozaki, K. (1994a) Characterization of two cDNAs that encode MAP kinase homologues in *Arabidopsis thaliana* and analysis of the possible role of auxin in activating such kinase activities in cultured cells. *Plant J.* **5**, 111–122.

Mizoguchi, T., Hayashida, N., Yamaguchi-Shinozaki, K., Kamada, H. and Shinozaki, K. (1994b) Characterization of a gene family of MAP kinases in *Arabidopsis thaliana*. Poster abstract no. 980, Fourth International Congress of Plant Molecular Biology, Amsterdam, 1994.

Neiman, A.M. (1993) Conservation and reiteration of a kinase cascade. *Trends Genet.* **9**, 390–394.

Pay, A., Jonak, C., Bögre, L., Meskiene, I., Mairinger, T., Szalay, A., Heberle-Bors, E. and Hirt, H. (1993) The Msk family of alfalfa protein kinase genes encodes homologues of shaggy/glycogen synthase kinase-3 and shows differential expression patterns in plant organs and development. *Plant J.* **3**, 847–856.

Perrimon, N. (1994) The genetic basis of patterned baldness in *Drosophila. Cell* **76**, 781–784.

Puziss, J.W., Hardy, T.A., Johnson, R.B., Roach, P.J. and Hieter, P. (1994) MDS, a dosage suppressor of an mck1 mutant, encodes a putative yeast homolog of glycogen synthase kinase 3. *Mol. Cell. Biol.* **14**, 831–839.

Roe, J.L., Rivin, C.J., Sessions, R.A., Feldmann, K.A. and Zambryski, P.C. (1993) The *Tousled* gene in *A. thaliana* encodes a protein kinase homolog that is required for leaf and flower development. *Cell* **75**, 939–950.

Ruel, L., Pantesco, V., Lutz, Y., Simpson, P. and Bourouis, M. (1993) Functional significance of a family of protein kinases encoded at the *shaggy* locus in *Drosophila. EMBO J.* **12**, 1657–1669.

Sagata, N., Daar, I., Oskaisson, M., Showalter, S.D. and Vande Woude, G.F. (1989) The product of the c-mos proto-oncogene as a candidate 'initiator' for oocyte maturation. *Science* **245**, 643–645.

Shibata, W., Banno, H., Hirano, Y.I.K., Irie, K., Machida, S.U.C. and Machida, Y. (1995) A tobacco protein kinase, NPK2, has a domain homologous to a domain found in activators of mitogen-activated protein kinases (MAPKKs). *Mol. Gen. Genet.* **246**, 401–410.

Siegfried, E., Perkins, L.A., Capaci, T.M. and Perrimon, N. (1990) Putative protein kinase product of the *Drosophila* segment-polarity gene zeste-white 3. *Nature* **345**, 825–829.

Stafstrom, J.P., Altschuler, M. and Anderson, D.H. (1993) Molecular cloning and exppression of a MAP kinase homologue from pea. *Plant Mol. Biol.* **22**, 83–90.

Tregear, J.W., Decroocq-Ferrant, V., Thomas, M., Bianchi, M.W., Le Guen, L., Lessard, P., Lecharny, A. and Kreis, M. (1995) Plant protein phosphorylation in signal transduction and the control of development: cloning and characterization of protein kinase genes from *Arabidopsis* and *Petunia*. In: *Protein Phosphorylation in Plants* (eds P.R. Shewry, N. Halford and R. Hooley). Oxford University Press, Oxford, in press.

Walker, J.C. (1993). Receptor-like protein kinase genes of *Arabidopsis thaliana. Plant J.* **3**, 451–456.

Walker, J.C. and Zhang, R. (1990) Relationship of a putative receptor protein kinase from maize to the S-locus glycoproteins of *Brassica. Nature* **345**, 743–746.

Wilson, C., Eller, N., Gartner, A., Vicente, O. and Heberle-Bors, E. (1993) Isolation and characterization of a tobacco cDNA clone encoding a putative MAP kinase. *Plant Mol. Biol.* **23**, 543–551.

Woodgett, J.R. (1990) Molecular cloning and expression of glycogen synthase kinase-3/factor A. *EMBO J.* **9**, 2431–2438.

Woodgett, J.R. (1991) A common denominator linking glycogen metabolism, nuclear oncogenes and development. *Trends Biochem. Sci.* **16**, 177–181.

Maize embryogenesis mutants

J.K. Clark

6.1 Introduction

As is the case with animal embryos, the plant embryo originates as a single-celled zygote which develops by directed cell division and subsequent differentiation into an embryo possessing the rudiments of the organs and tissue types of the adult body. The plant embryo consists of three structures:

(i)   the *embryonic axis*, characterized by shoot and root meristems at either end, which will grow into the adult plant;
(ii) one or more modified leaf-like storage organs, the *cotyledons*, which nourish the embryo during germination;
(iii) the *suspensor*, which serves to anchor and nourish the young embryo during its early development.

Although each of these structures develops to a different extent in different plant groups, all plant embryos have the form of a presumptive embryonic axis flanked by its cotyledons, together termed the *embryo proper*, situated above an elongated suspensor. The regularity of the developmental sequences reflects underlying genetic programmes, while similarities in embryo structure and ontogeny across wide taxonomic distances attest to their evolutionary conservation. For comprehensive reviews of plant embryogenesis see Maheshwari (1950), Johri (1984) and Raghavan (1986), and for reviews of maize embryogenesis see Sheridan and Clark (1993a).

The mature grass embryo consists of a well-differentiated embryonic axis surrounded by a single massive cotyledon, the scutellum (Rost and Lersten, 1973). Maize mutants have been an especially favourable system for identifying and studying individual genes important for morphogenesis of the grass embryo. Maize embryogenesis has been well characterized, a wealth of genetic resources has been developed to aid mapping and cloning, and over many years a large number of mutants have been identified. Early collections of spontaneous mutants clearly demonstrated that defects in single genes could perturb embryogenesis. Their phenotypic diversity provided evidence that mutant analysis would be a fruitful experimental approach. These collections defined three important classes of mutants: *defective kernel* (*de*, *dek*) mutants, in which both embryo and endosperm are affected, *embryo-specific* (*gm*, *emb*) mutants, in which the embryo but not the starchy endosperm is affected and *viviparous* (*vp*) mutants, in which maturation is affected. The *vp* mutants are considered by McCarty in Chapter 8. This review will focus on the maize *dek* and *emb* mutants. The course of embryogenesis in maize will be described first, then the *dek* and *emb* mutants and their contribution to our understanding of embryo morphogenesis will be discussed, and finally their potential for future studies of embryogenesis and evolution will be considered.

6.2 Maize embryo development

During embryogenesis, the maize embryo increases in length by 100-fold and develops from a single-celled zygote into a well-differentiated miniature plant (Figures 6.1 and 6.2). The classic comprehensive study of maize embryogeny is the histological investigation reported by Randolph (1936). Kiesselbach (1949) and Sass (1977) have also provided excellent descriptions of the course of maize embryogenesis, based on sectioned materials. While these detailed descriptions are essential, a system of embryo stages is also necessary in order to provide a conceptual framework for the complex continuum of ontogeny. Wang (1947) divided embryogeny into six stages, and Paxson (1963) defined 11 stages of embryogenesis based on the differentiation of the vascular system. These schemes are of limited use because they do not encompass the entire developmental sequence in detail, and they cannot be applied to both dissected and sectioned materials. Abbé and Stein (1954) compared the external features of the embryo with its state of internal differentiation, and devised a staging system based on the development of leaf primordia. Their system has been widely adopted. Maize embryogeny has been divided into three phases, the first consisting of early regionalization and establishment of polarity, the second consisting of the rapid morphogenesis of the major regions and structures of the embryo, and the third consisting of the period of maturation (Clark and Sheridan, 1991; Sheridan and Neuffer, 1981). In the following description, 'days after pollination (DAP)' refers to maize grown under field conditions in Grand Forks, North Dakota, USA. Embryo stages are those of Abbé and Stein (1954).

6.2.1 *Polarization period: establishment of the embryo proper and suspensor stage (0–7 DAP)*

During this period the embryo elongates and differentiates into an upper embryo proper and a lower suspensor. This defines the subsequent overall polarity of the embryo, which is congruent with that of the egg and embryo sac established in the gametophyte.

Zygote to proembryo stage (Figures 6.1a and 6.2a). Fertilization occurs at about 15–24 h after pollination, and cell division in the zygote begins a day or two later (Diboll, 1968; Diboll and Larson, 1966; VanLammeren, 1986, 1987). The initial division is not precisely positioned but is invariably transverse to the axis of the embryo sac. The upper cell ultimately gives rise to the embryo proper, and the lower cell gives rise to the suspensor. The first division of the upper cell is always longitudinal. The upper region of the developing embryo divides more rapidly than the lower region, producing a club-shaped proembryo containing small dense cells in its upper region and larger vacuolated cells in its lower region. Cell division in the proembryo does not follow a highly ordered pattern, and the boundary between the two regions of the proembryo is not distinct. This stage is roughly equivalent to the globular stage of dicot embryos. By about 5 DAP the proembryo has reached a length of 0.2 to 0.4 mm.

Transition stage (Figures 6.1b and 6.2b). The first evidence of differentiation in the embryo is the appearance of an epidermis on the club-shaped proembryo. During the following 2 days the upper region develops into a distinct ball of small dense cells, the embryo proper, while the lower region continues to elongate to form a large-celled suspensor. The embryo is radially symmetrical in overall form, but not in internal cell patterns. By this time the embryo, including the suspensor, is about 1.0–1.5 mm in length.

6.2.2 *Morphogenesis period: differentiation of the embryonic axis and scutellum (7–11 DAP)*

This consists of the period between the mid-transition stage and stage 1. During this stage the embryo first manifests bilateral symmetry, establishes a new structural (and perhaps functional) axis, and initiates all of the structures and tissue types present in the adult vegetative plant.

Late transition stage (Figures 6.1b and 6.2b). Following the appearance of the epidermis the face of the embryo becomes flat, marking the shift from radial to bilateral symmetry. An internal cone-shaped group of meristematic cells appears adjacent to the flattened surface. This primary meristematic region will eventually form the shoot–root axis of the mature embryo, defining the body of the adult plant and establishing a new axis within the embryo axis; this axis is not congruent with that of the original proembryo. The early establishment of bilateral symmetry is demonstrated by the conclusion, based on clonal analysis of sectors induced by X-irradiation of embryos shortly after pollination, that the shoot apical meristem is derived from a group of 2–3 initials whose developmental fate is determined in the transition stage (Poethig *et al.*, 1986). Following establishment of the primary meristematic region, the remainder of the embryo proper continues to expand upward and to widen and deepen at its base. Expansion of the bulk of the embryo proper into the parenchymatous scutellum continues throughout embryogenesis. The remainder of the transition-stage embryo proper eventually

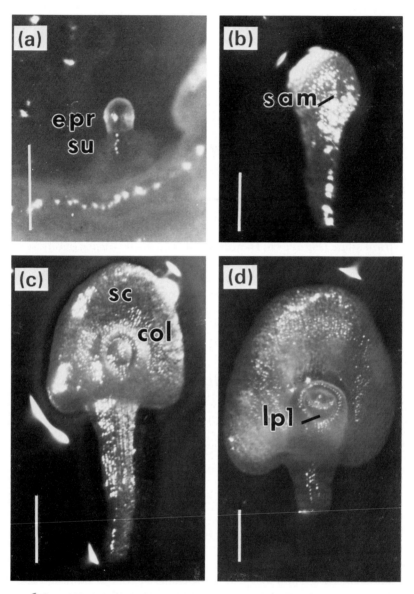

Figure 6.1. *Maize embryo morphogenesis. Dissected embryos from the proembryo stage to stage 1 in face view. New structures are indicated for each stage. (a) Proembryo. The club-shaped embryo consists of a dome-shaped embryo proper (epr) above a thick suspensor (su). (b) Transition stage. The shoot apical meristem (sam) is visible as a dimple on the flattened face of the embryo. (c) Coleoptilar stage. The coleoptilar ring (col) surrounds the shoot apical meristem. The remainder of the embryo proper has expanded into the scutellum (sc). (d) Stage 1. The first leaf primordium (lp1) is growing upward across the shoot apical meristem. Scale bar = 0.5 mm.*

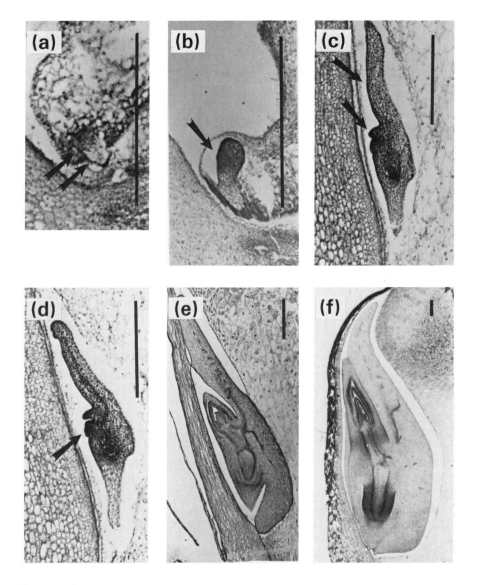

Figure 6.2. *Stages of maize embryogenesis. Median longitudinal (sagittal) sections. The face of the kernel bearing the silk scar is at the left. Arrows indicate structures described in Figure 6.1. (a) Proembryo, with embryo proper (upper arrow) and suspensor (lower arrow). (b) Transition stage, with the cone-shaped region of meristematic cells that will become the shoot apical meristem. (c) Coleoptilar stage, with scutellum (upper arrow) and coleoptilar ring (lower arrow). (d) Stage 1, with the first leaf primordium growing upward over the shoot apical meristem. (e) Stage 3. Note that the embryonic axis constitutes a well-differentiated miniature plant. (f) Stage 6, mature embryo. Note the extensive enlargement of the scutellum, and the elongation of the mesocotyl region between the base of the shoot apex and the point of exit of the provascular strands into the scutellum. Scale bar = 0.5 mm.*

forms the coleorhiza, a sheath of parenchymatous tissue surrounding the embryonic axis below the coleoptile.

Coleoptilar stage (Figures 6.1c and 6.2c). During the following 2 days the embryonic axis is established, oblique to the axis of the transition-stage embryo. The shoot apex is apparent as a bulge on the anterior face of the embryo, and the root apex is evident as a dark-staining region within the embryo. In the space between them, at the site of the future scutellar node, a strand of vascular procambium extends into the scutellum. The coleoptilar ring appears as a bulge on the face of the scutellum encircling the shoot apex. The scutellum expands rapidly, and at this time, about 9 DAP, the embryo, including the suspensor, has reached a length of about 1.5–2.0 mm.

Stage 1 (Figures 6.1d and 6.2d). Shortly after the coleoptilar stage has been reached, the first leaf primordium is initiated on the front surface of the shoot apex. It grows upward to cover the face of the shoot apex, eventually folding behind it. Thus by the end of the morphogenesis period, the three key regions of the embryo have differentiated, and all of the tissue types and structures of the vegetative plant have made their initial appearance.

6.2.3 *Maturation: elaboration of structures, maturation and dormancy (11–45 DAP)*

During this period the embryonic axis differentiates into a miniature plant, storage products are deposited in the scutellum, and the rate of development slows down as dormancy is imposed upon the embryo.

Stages 2–4 (Figure 6.2e). By the time the second leaf primordium has appeared opposite the first, the coleoptilar ring has developed into the coleoptile, ensheathing the developing shoot axis except for a small pore. During stage 2, the scutellum becomes opaque as storage proteins are deposited within it. By the time the third leaf primordium has appeared, the suspensor has disappeared. During the maturation period the scutellum grows rapidly to form a large shield-shaped structure composed of parenchyma cells. It eventually overgrows most of the shoot axis, starting at about the time of the appearance of the fourth leaf primordium (stage 4).

Stages 5–6 (Figure 6.2f). At kernel maturity, about 45 DAP, the embryo has differentiated to form five or six leaf primordia, and the epidermis of the scutellum has undergone specialized differentiation in two regions (Wolf *et al.*, 1952). On the side facing the endosperm, it has undergone additional anticlinal divisions to produce the scutellar epithelium, while on the side adjacent to the coleoptile its walls have developed a heavy cuticle (Wolf *et al.*, 1952). The embryonic provascular system extends from the scutellar node upward and downward into the scutellum, and provascular strands are present in the embryonic axis and leaf primordia. A few vascular cells at the tip of the scutellum have developed secondary wall thickenings (Paxson, 1963). The primary root of the embryonic axis is well developed, and at the base of the shoot apex one or two seminal roots have differentiated at the coleoptilar node. The primary root is ensheathed by the coleorhiza, a structure composed of parenchymatous cells derived from the lower region of the transition-stage embryo proper. During the last half of the maturation phase the region of the embryonic axis between the coleoptilar and scutellar nodes elongates by cell division to form an extended mesocotyl, lifting the shoot apex well above the scutellar node. The shoot apical meristem is structurally undifferentiated at kernel maturity. However, experiments utilizing radiation-induced pigmentation chimeras have shown that the meristem in the mature embryo is in fact already organized into stacked regions that will produce all of the successive nodes of the mature plant (Johri and Coe, 1982). At kernel maturity the embryo has reached a length of 7.0–10.0 mm.

The endosperm of maize is composed of three morphological regions: the starchy endosperm, the basal transfer cells and the aleurone (Kiesselbach and Walker, 1952). The starchy endosperm starts to undergo cell division and rapid expansion before the zygote divides. The maize aleurone is a single cell layer which begins to differentiate when the embryo is at the transition stage (Kyle and Styles, 1977). Aleurone cells divide periclinally during the morphogenesis period and anticlinally thereafter (Randolph, 1936). In barley, the embryo and aleurone co-express many genes (see Chapter 10, Brown *et al*.), and many maize embryogenesis mutants also affect aleurone morphogenesis (see below).

6.3 The *defective kernel* (*dek*) mutants

Defective kernel (*dek*) mutations result in defects in the development of both the endosperm and the embryo. Recessive *defective* (*de*) seed mutants were first described by Jones (1920) in self-pollinated ears from a variety of agronomic maize materials. These mutants were lethal in that they were incapable of germination. Mutant kernels generally displayed a severely collapsed endosperm and a tiny embryo at maturity. It is clear from the variation in severity of the *de* mutants and from the heterogeneity of their backgrounds that the collection probably represented many different gene loci. Manglesdorf (1923, 1926) used genetic and histological techniques to study the inheritance and phenotype of 14 spontaneous *de* mutations, designated *de1* to *de14*. All were recessive, and only one case of allelism between two of the mutants was found. Embryos were blocked at characteristic stages from before primordia formation to late in development. The 14 mutations formed a graded series with regard to the progress of embryo differentiation. At least one displayed morphological abnormalities, and embryos of another were present during early embryogenesis but had degenerated and disappeared by kernel maturity. Manglesdorf recognized these mutants and others as belonging to several classes, representing arrest at sequential stages of a developmental programme. Mutants devoid of both endosperm and embryo, mutants lacking embryos, and mutants with a reduced endosperm were interpreted as resulting from defective fertilization. Defects at later stages of development were thought to produce seeds incapable of growth and germination. Mutants that displayed impaired dormancy were also recognized. Although all of these mutants have been lost, the early work of Jones and Manglesdorf demonstrated that *dek* mutants are among the most common of mutations, and established that mutations in single genes can perturb the course of maize embryogenesis. Manglesdorf's vision of embryogenesis as an unfolding developmental programme, and his appreciation of the value of defective kernel mutants in defining its components, was ahead of its time.

Emerson (1932) described a *zygotic lethal* (*zl1*) mutation from a Peruvian stock. No defective kernels were observed, but the presence of a recessive zygotic lethal was inferred from the reduced transmission of markers conferring pericarp colour and male sterility on the chromosome carrying the putative mutant, as well as a reduced seed set on self-pollinated ears. This mutation was precisely mapped at 1.5 centimorgans proximal to the *P1* locus on chromosome *1S* (Emerson, 1939), making it very tightly linked to the *dek1* gene (described below). Coulter (1925) postulated the presence of a zygotic (designated *l1*) lethal on *9S*, based on aberrant aleurone colour ratios in crosses involving *c1*. Brink (1927) described a new lethal *de* mutant, *de15*, also on *9S*, which mapped close to *l1* but was not allelic to it. The complete failure of both embryo and endosperm development in the two zygotic lethals reported by Emerson and Coulter implies that these genes act immediately after fertilization. These mutants have been lost. However reports demonstrate that screening for mutations of this class can be achieved by monitoring the transmission frequency of viable markers along each of the chromosome arms. Since the maize map is well covered by morphological and RFLP markers, intensive screening of the entire genome is feasible.

In a study of mutant induction by ultraviolet (UV) radiation, Sprague (1941) recovered eight recessive lethal *aborted seed* (*ab*) mutations among 71 non-viable and 24 viable mutations. The *ab* mutant kernels consisted of a collapsed, empty pericarp containing a tiny endosperm and embryo. This phenotype, illustrated in Figure 12 of Jones (1920), is very common among *dek* mutations. Three of six *ab* mutants studied in the F_3 generation by Sprague showed a reduced ratio of mutant kernels on self-pollinated ears, indicating a reduced sexual transmission of the mutant allele. This is a common feature of *dek* and *emb* mutations, and raises the possibility that these genes are also expressed in the gametophyte (Table 6.1; Clark and Sheridan, 1988). The UV mutant collection, which no longer survives, is of value in demonstrating that a variety of mutagenesis strategies may be used to induce mutations in the *dek* genes.

Neuffer has isolated 207 *defective kernel* (*dek*) mutants in a collection of 2457 recessive mutations induced by EMS mutagenesis (Neuffer and Sheridan, 1980, and this work has been summarized by Sheridan and Neuffer (1981, 1986) and Neuffer *et al.* (1986). The mutants were first named using laboratory designations referring to their endosperm phenotype (Neuffer and Sheridan, 1980). To date, *dek* mutants have been found on 18 of the 19 arms that can be tested with the B-A translocation stocks (M. G. Neuffer, personal communication). Selected

Table 6.1. *Maize* embryo-specific (emb) *mutations*

Mutant	Stage	Segregation		Arm	Germination	
		No. of kernels	Percentage of mutants		No. of kernels	No. of germination events
emb*-8501	c–3	100	24	4L	109	5
emb*-8502[a]						
= emb4	t–c	130	24	1S	100	1
emb*-8503[a]						
= emb4	t–c	120	20	1S	100	3
emb*-8504	t–c	214	22	5S	16	0
emb*-8505	t–c	214	22	nt	nt	
emb*-8506	c	310	25	1S	nt	–
emb*-8507	2–4	94	19	nl	100	7
emb*-8508	c–1	224	23	nl	70	5
emb*-8509	t–c	100	29	4L	80	5
emb*-8510	c–2	117	23	nl	75	2
emb*-8511	t–c	124	23	nt	nt	–
emb*-8512	t	75	21	3L	100	3
emb*-8513						
= emb11	3–6	103	12***	4L	35	2
emb*-8514						
= emb8	c–2	150	23	4L	100	6
emb*-8515	t–c	104	20	3L	35	0
emb*-8516	p–t	212	28	nt	nt	–
emb*-8517						
= emb12	t–c	51	23	1S	45	2
emb*-8518						
= emb5	t–c	67	18	2L	45	2
emb*-8519[b]						
= emb19	t–5	93	19	1S	50	2
emb*-8520[b]						
= emb20	t–2	183	17**	1S	103	8
emb*-8521						
= emb9	t–2	100	13**	3L	100	2
emb*-8522						
= emb2	t	140	21	9S	105	5
emb*-8523	p–c	180	14***	nl	40	0
emb*-8524	p–c	187	18*	nl	nt	–
emb*-8525	p	150	11***	nl	44	1
emb*-8526	c–2	160	13***	nt	nt	–
emb*-8527	c	133	20	nl	75	1
emb*-8528	t–c	99	19	nl	15	0
emb*-8529	p t	78	14*	nl	95	3
emb*-8530	p–t	157	10***	nt	nt	–

Continued overleaf

Table 6.1. *Continued*

Mutant	Stage	Segregation			Germination	
		No. of kernels	Percentage of mutants	Arm	No. of kernels	No. of germination events
emb*-8531						
= *emb1*	z–c	118	28	1S	100	1
emb*-8532	c–4	206	23	3L	25	0
emb*-8533	c	105	14**	nl	20	0
emb*-8534	t–1	396	22	4L	100	2
emb*-8535	p–c	141	21	nl	50	0
emb*-8536[c]	c	166	22	nl	nt	–
emb*-8537[c]	t–c	167	19	4L	20	1
emb*-8538[c]	t–c	149	24	4L	30	2
emb*-8539[c]	t–c	139	28	nl	50	2
emb*-8540						
= *emb6*	c	197	18*	4L	50	0
emb*-8541	c–2	48	27	nt	nt	–
emb*-8542	c	184	19	nl	60	15
emb*-8543	t	123	20	nl	65	4
emb*-8544	t–1	186	21	nl	25	5
emb*-8545						
= *emb10*	2–6	64	23	1S	75	8
emb*-8546	p–t	163	26	nl	100	4
emb*-8547						
= *emb3*	t	111	23	4L	100	0
emb*-8548	p–c	190	26	nt	100	0
emb*-8549						
= *emb13*	t–2	86	23	1S	100	1
emb*-8550	2–6	194	15**	1S	100	3
emb*-8551	t–c	120	22	3L	10	5

[a, b, c]May be a single mutation event, based on pedigree.
nt = not tested; nl = at least partially tested but not located.
*Significantly different from 25% at $P = 0.05$; **Significantly different from 25% at $P = 0.01$; ***Significantly different from 25% at $P = 0.001$.
Compiled from Clark and Sheridan (1991), Sheridan and Clark (1992) and Clark (unpublished results).

mutants whose chromosome arm location has been determined have been assigned the formal gene designations *dek1* to *dek33 (Neuffer et al.,* in press). These non-allelic *deks* define 33 different *dek* loci. The rate of allelism among mutants on the same chromosome arm is low. On the basis of these data, it was estimated that at least 250 genes are involved in embryo development (Neuffer *et al.,* 1986).

Most of the EMS *dek* mutants are lethal at kernel maturity. The collection was screened for auxotrophy by culturing immature embryos on basal and enriched media. This led to the operational definition of three classes of mutants (Sheridan and Neuffer, 1980, 1981, 1982). At least 108 mutants are nutritionally conditional lethals, with normal embryo morphology and the ability to germinate precociously when cultured as immature embryos. These mutants are likely to have a defect in biochemistry or metabolism. Four are proline auxotrophs that are allelic to the *pro-1* mutant of Gavazzi *et al.* (1975). A second class, containing four mutants, consists of non-germinating mutants. These had normal embryos which did not germinate when cultured. The third class, the developmental *dek* mutants, is composed of 17 mutations which are lethal and morphologically abnormal. Four of these constitute an allelic series at the *dek1* locus on *1S*. Seven others have been located at six other chromosome arms (Sheridan and Neuffer, 1982).

Analysis of the non-germinating and developmental *deks* has provided information about the role of *dek* genes in embryogenesis. The terminal phenotype of the non-germinating and developmental mutants was examined in dissection (Sheridan and Neuffer, 1982, 1986). The non-germinating mutants were blocked during the mid-maturation period (stages 2–3), but were morphologically normal, indicating that they may be involved in determining the rate or extent of maturation and dormancy. The developmental *dek* mutants were morphologically abnormal. They were blocked during all three periods of embryogenesis, but most were blocked prior to or during the morphogenesis period. The four *dek1* alleles were blocked during the polarization period, and 10 other mutants were blocked during the morphogenesis period. Two mutants were blocked in the early maturation period (stages 1–2), and were grossly deformed.

When the course of development of a number of the developmental *dek* mutants was examined in sectioned materials, all proved to be retarded throughout development, indicating that these genes act at a very early stage. They varied in the point at which they departed from the course of normal embryogenesis and in the nature of their ensuing abnormal morphogenesis (Clark and Sheridan, 1986, 1988; Sheridan and Clark, 1987; Sheridan and Thorstenson, 1986). One mutant, *dek22*, displayed a profound retardation of otherwise normal development. Its embryos simply stopped growth and development at the transition stage. Three mutants displayed impaired shoot-meristem morphogenesis. Embryos of *dek1* formed a large proembryo-like mass containing a root meristem but no shoot meristem (Sheridan and Neuffer, 1981). Embryos of *dek23* reached the coleoptilar stage with a root meristem and elongated scutellum. However, no shoot meristem was formed, and its site on the face of the embryo underwent selective necrosis, followed by necrosis of the entire embryo. Embryos of *cp*-1418* also reached the coleoptilar stage. They formed an embryonic axis, but the shoot meristem later disappeared, although the root meristem was retained and slow growth of the embryo continued. Four mutants suffered a breakdown of organization and underwent proliferative growth after varying periods of normal morphogenesis. Embryos of *rgh*-1210* reached the proembryo stage, but by the transition stage they had developed meristematic knobs, and thereafter they displayed growth by disorganized cell division and cell enlargement. Embryos of *dek31* reached the transition stage with a primary meristem, but did not differentiate an embryonic axis or scutellum, and continued to grow by cell enlargement and vacuolation to form a large mass of hypertrophied cells. Embryos of *bno*-747B* reached the coleoptilar stage with an embryonic axis and scutellum, but subsequently the identity of the entire embryonic axis was lost, and growth continued by cell enlargement and vacuolation. In the mutant *fl*-1253B*, development of the scutellum became uncoupled from that of the embryonic axis. After reaching a relatively normal stage 1, the scutellum ceased growth and development and its cells became enlarged and vacuolated, while the coleoptile and coleorhiza proliferated by cell division, forming a giant structure composed of small cells ensheathing the embryonic axis. In these four mutants the aleurone also underwent proliferative growth.

Seven of the nine developmental *dek* mutants that were examined in detail showed a reduced frequency of mutant kernels on self-pollinated ears, suggesting that these mutants may be expressed in the gametophyte generation (Sheridan and Clark, 1993a). In the case of *fl*-1253B*, progeny ratios in reciprocal crosses and examination of the distribution and frequency of mutant kernels on self-pollinated ears revealed that male transmission was significantly reduced, that the defect involved pollen viability or germination but not tube growth, and that the observed variation in the segregation ratio was due to variable expressivity of the *fl*-1253B* gene, rather than to a second factor suppressing transmission (Clark and Sheridan, 1988).

The collection of EMS *deks*, which has been deposited in the Maize Stock Center, represents the largest and most intensively studied group of *dek* mutants to date. It continues to be a valuable resource for the genetic dissection of maize embryogenesis.

Recently, a collection of 63 Robertson's *Mutator*-derived *deks* has been described (Scanlon *et al.*, 1994). At least 14 of these *deks* represent previously undescribed *dek* loci. They are located throughout the genome; 53 *dek* loci have been located among 15 chromosome arms, and 21 have been mapped more precisely. They differ greatly in embryo and endosperm phenotype and viability. The most common phenotypes observed were *reduced endosperm* (*ren*) and *empty pericarp* (*emp*). Embryos in the *emp* and many of the *ren* mutants were severely defective (*gm*). Of the 63 *deks*, 31 are embryonic lethals, and 29 of these have very poorly developed embryos, or none at all, at maturity. Ten are seedling lethals that die before the five-leaf stage, and most of the remainder germinate only at very low levels. Three cases of allelism between the new *dek* loci have been found, at *brn1*, *dsc1* and *ren2*. All of these mutants are full or partial lethals with defective embryos. One of them, *brn1*, shows allelic variation in expression. The *brn1-R* allele is a seedling lethal, while the *brn1-3071* allele is an embryo lethal. This difference is maintained across many genetic backgrounds (J.K. Clark, unpublished observations). Nine of the mutants in this collection are allelic to five of the EMS *dek* loci; *dek1*, *dek5*, *dek7*, *dek25* and *dek31* are represented in both collections. Because these mutants are potentially transposon tagged, this large collection of *Mutator*-derived *deks* offers the potential for cloning *dek* genes involved in morphogenesis of the embryo.

Analysis of embryo–endosperm interactions in the EMS *dek* genes and in the *brn1* alleles using B-A translocations revealed that the mutant phenotype of the embryo and of the endosperm results from the genotype of each, and is not simply a secondary effect of the developmental state of the other tissue (Clark and Sheridan, 1986; Neuffer *et al.*, 1986; Sheridan and Neuffer, 1982; J.K. Clark, D.S. Robertson and P.S. Stinard, unpublished results). It is clear that *dek* genes play important roles in embryo morphogenesis, and they should not be dismissed on account of their pleiotropy as merely 'housekeeping' genes.

6.4 The *embryo-specific* (*emb*) mutants

Embryo-specific (*emb*) mutations result in defects in the development of the embryo, while development of the endosperm is essentially normal. Thus they may define genes whose products are needed specifically for embryo morphogenesis. The *emb* phenotype consists of a kernel which has a well-filled and apparently normal endosperm, but a sunken region on its adaxial face marking the site of a reduced or absent embryo. This phenotype was described and termed *germless* (*gm*) by Demerec (1923), who reported *gm* mutants from self-pollinated ears of commercial varieties of corn. Segregation ratios of 3:1, 15:1, 63:1 and 9:7 occurred among these ears and their progeny. On the basis of these findings, Demerec estimated that a minumum of four genes are involved in producing the *gm* phenotype, and that three of these constitute a triplicate factor. He concluded that *gm* mutants are very common, based on his observation that self-pollinated ears of 75% of the commercial varieties he examined as well as five inbreds examined by Wiggens segregated for *gm* kernels. Similarly Goodsell (1927), cited by Wentz (1930), observed that self-pollinated ears of 15 of 19 commercial strains involved in a yield trial produced *gm* kernels.

Wentz (1930) studied the transmission of *gm* mutants from 23 named maize varieties, and he found their inheritance to be complex. He obtained only two stocks that segregated consistently in a 3:1 ratio from among 52 lines tested. Progenies of most stocks were highly variable in their segregation ratio, even though the parent ears segregated in a 3:1 ratio. A low frequency of kernels with the *emb* phenotype occurs on genetically normal ears as a result of physiological accident (Manglesdorf, 1926). In order to assess whether this phenomenon could account for the lowest frequencies, Wentz examined nine parental lines segregating 0.3–12% *gm* kernels. He found a strong correlation between segregation frequencies of parent and progeny ears, ruling out this explanation. A second factor suppressing transmission was shown to occur in advanced generations of *gm2*. However, the majority of deviations could not be attributed to this phenomenon, and remain unexplained. Alternative possibilities not tested by Wentz are that the observed variability in segregation ratios could be due either to additional *gm* mutations arising in the stock, which would be impossible to detect by kernel phenotype, or to inherent variable expressivity of the *gm* mutant itself (Clark and Sheridan, 1988). Subsequently, 30 heritable *gm* mutants were recovered among 71 lethal UV-induced mutations (Sprague, 1941). These also showed aberrant segregation ratios on self-pollinated ears; 20 mutants showed a reduced frequency, with less than 20% mutant kernels instead of the expected 25%, and two had an excess of mutant kernels, with 36% and 40% mutant kernels.

The *gm* mutants have been lost. However, these early reports define the phenotype and show that *emb* mutations are very common. They also provide a cautionary note as to the difficulties in handling mutants of these developmentally interesting genes.

Embryo development in a spontaneous *emb* mutant, *GermlessS*, was examined in sectioned materials (Sass and Sprague, 1950). Mutant embryos became retarded at the transition stage. They reached the coleoptilar stage with an embryonic axis, but subsequently the identity of the embryonic axis was lost as the coleorhiza, coleoptile and leaf primordia underwent enlargement by cell division. Simultaneously, the scutellum became hypertrophied and necrotic, and eventually the embryo degenerated so that the mature kernel contained only its debris. The aleurone underwent proliferation, but otherwise endosperm development was normal. The course of development of *GermlessS*, which is reminiscent of the proliferative EMS *dek* mutants discussed above, confirms that *emb* genes play an important role in embryo morphogenesis.

Three additional mutations were first described as seedling mutations, but their phenotypes are clearly rooted in abnormal morphogenesis of the embryo. The *accessory blade* mutation (Sass and Sprague, 1949) produced adventitious leaves and fusions between the embryonic leaves, coleoptile and scutellum. The authors concluded that this was due to excessive meristematic activity at the edges of these structures. Examination of mutant embryos early in development revealed that they were retarded by the transition stage, and that the upper scutellum, coleorhiza, coleoptile and embryonic leaves showed excess thickening at stage 1 (Sass and Sprague, 1950). A second mutant, *siamensis*, was recovered twice from Brazilian lines (Horovitz and Sanguineti, 1936). Mutant kernels produced a variety of twinned embryonic axis structures, ranging from entire twinned seedlings joined at the scutellar node to twinned shoots emerging from a common coleoptile. The variation in twinning phenotype presumably occurred as a result of bifurcation of the embryo at different stages of embryonic axis formation. A twinning mutation, *Growing point* (*Gp*), was also recovered among the 95 heritable UV-induced mutations (Sprague, 1941). Its phenotype was like that of *siamensis*; mutant seedlings contained up to four plumules and radicles. Sprague interpreted its complex pattern of transmission as indicating that *Gp* may be a dominant factor conditioned by two or more genes. Although all three of these mutations have been lost, their phenotypes are of compelling interest. Mutants with similar phenotypes should be sought during the screening of modern mutant collections.

A total of 51 recessive *embryo-specific* (*emb*) mutations, representing at least 45 different mutation events, have been recovered from self-pollinated ears of active Robertson's *Mutator* stocks (Table 6.1; Clark and Sheridan, 1991; Sheridan and Clark, 1993b). The *emb* designation has been adopted in order to avoid confusion with the use of the term 'germless' *gm* to refer to *dek* mutants which have substantially reduced embryos (Scanlon *et al.*, 1994). Selected *emb* mutants which have been shown to be located on a particular chromosome arm have been assigned the formal gene designations *emb1* to *emb13* (Table 6.1) (Neuffer *et al.*, in press). To date, 27 mutants have been located on a chromosome arm by the B-A translocation stocks. Single mutants are located on *2L*, *5S* and *9S*; clusters of five mutants are located on *3L*, nine mutants (representing eight independent events) on *4L*, and 10 mutants (representing eight independent events) on *1S*. Allelism tests among seven of the *1S* mutants defined six complementation groups (J.K. Clark, unpublished results). These initial findings suggest that *emb* genes, like *dek* genes, are abundant, and that they are likely to be found throughout the genome. Most of the *emb* mutants are lethal (Table 6.1). The low level of germination observed in the mutants with early blocks is probably due to the inclusion of wrongly scored normal kernels in the sample, since the kernel phenotype is more difficult to score than that of *dek* mutants. However, some mutants blocked during the maturation period may be viable at low levels. Twelve *emb* mutants showed a significantly reduced frequency of mutant kernels on self-pollinated ears (Table 6.1). This is consistent with earlier observations of *emb* and *dek* mutants, and suggests that these *emb* genes may function in the gametophyte generation.

The mutant embryo phenotype of all 51 *emb* mutants was determined by examining mature mutant embryos in fresh dissection (Clark and Sheridan, 1991; Sheridan and Clark, 1993b). The *emb* mutants are blocked during all three periods of embryogenesis (Tables 6.1 and 6.2). A total of 41 mutants (80%) were blocked during the first two periods of embryogenesis, when the major morphogenetic events take place. This is in contrast to the EMS *dek* mutants, where only 11 mutants (5%) were blocked during these periods. Embryos of 12 mutants were blocked during the period of polarity establishment, from the proembryo to the mid-transition stage. Embryos of another 29 mutants were blocked during the morphogenesis period, from the mid-transition stage to stage 1. This is significant in that the morphogenesis period occupies only about 5 days of the 45-day embryogenesis span. Embryos of 10 mutants were blocked during the maturation period.

Most of the *emb* mutants are morphologically abnormal and have distinctive phenotypes (Clark and Sheridan, 1991; Sheridan and Clark, 1993b; Table 6.2). Embryos of one mutant, *emb*-8546*, were blocked at a very early stage with apparently normal morphology, as was the case with the EMS *dek22* mutant. Although these mutants could be associated with a metabolic defect, the presence of a well-formed endosperm in *emb*-8546* and the sharp stage-specificity in both mutants are more consistent with blocking of a process directly concerned with morphogenesis. In embryos of eight mutants blocked during the polarization period, death of the embryo proper occurred, with a phenotype similar to that of the *dek* mutant *cp*-1365* (Sheridan and Neuffer, 1982). In three of the *emb* mutants, failure of the embryo proper was accompanied by proliferation of the suspensor. Suspensor proliferation following death of the globular-stage embryo

Table 6.2. *Phenotypes of 51* Mutator-*derived* emb *mutants*

	Period of terminal block		
	Polarization (proembryo–transition)	Morphogenesis (transition–stage 1)	Maturation (stage 2–stage 6)
Normal morphology	8546		
Irregular but healthy embryo	8522 8530 8535		
Death of the embryo proper	8516* 8525* 8529* 8543 8549*		
Death of the embryo proper, proliferation of the suspensor	8512* 8531* 8547*		
Irregular scutellum		8509 8524 8515* 8548 8517* 8551 8520^b*	8507 8521*
Irregular scutellum, embryonic axis missing		8504* 8533* 8505 8534 8506 8536^c 8511 8537^c 8518 8538^c 8519^b 8539^c* 8523 8542 8527 8544* 8528*	8510* 8514 8526 8550^d
Irregular scutellum, displaced embryonic axis		8508 8540*	8501
Truncated scutellum		8541*	8545^d
Elongated wavy embryo		8502^a* 8503^a*	
Premature greening			8532
Small embryo, non-germinating			8513

[a,b,c]May be a single mutation event, based on pedigree.
[d]Allelic mutations.
*Embryos become necrotic.

proper has also been observed in *Arabidopsis* mutants, and has been attributed to the release of the suppressed developmental potential of the suspensor (Schwartz *et al.*, 1994). The occurrence of this mutant phenotype in taxonomically distant species suggests that the developmental mechanisms underlying the early steps of embryogenesis have been conserved during angiosperm evolution.

Irregular shape, indicating proliferative growth, was the predominant feature of the phenotype of three *emb* mutants blocked during the polarization period, seven mutants blocked during the morphogenesis period and two mutants blocked during the maturation period. Necrosis was observed in the embryos of four of these mutants. As in the case of the proliferating EMS *dek* mutants, these *emb* genes may be involved in the maintenance of organization rather than the differentiation of specific structures. In embryos of another 20 *emb* mutants the embryonic axis was absent at maturity, and in embryos of three others it was displaced. Proliferative growth occurred in all of these mutants, and necrosis was observed in the embryos of seven of them. Many of these mutants may be similar to the proliferative EMS *dek* mutants in which embryonic axis structures are formed but are obliterated before maturity. However, it is also likely that some of these *emb* mutants show impaired establishment or siting of the shoot apical meristem, and proliferative growth may be a secondary effect of the failure to differentiate an embryonic axis.

Four of the *emb* mutants show impairment of aspects of embryonic patterning. Embryos of *emb**-8541 and *emb**-8545 displayed alterations in the specification of a particular structure, namely the scutellum. Embryos of *emb**-8502 and *emb**-8503 (representing a single mutation event) elongated excessively, indicating that this gene is involved in shape determination. Because of the uniformity of the mutant phenotypes and the fact that these phenotypes have no counterparts among the EMS mutants examined to date, is is likely that the *emb* genes represented by this group of mutants are specifically involved in patterning processes in the embryo. Embryos of the mutant *emb**-8532 undergo premature greening and coleoptile extension, while those of *emb**-8513 cannot germinate. These mutants, like the non-germinating EMS *dek* mutants, are likely to display defects in the physiological processes of maturation and dormancy.

The six *emb* mutants initially identified on chromosome arm *1S* were selected for further study, and additional *1S emb* mutants were found among *Mutator* and *Ac/Ds* materials (Table 6.3). Complementation tests have revealed that they represent six non-allelic loci. The low level of allelism among mutations on this one arm provides more specific evidence for the existence of many *emb* genes in the maize genome.

The stage of terminal block and mature embryo phenotypes for the *1S emb* mutants was examined in more detail, using a large number of dissected embryos as well as sectioned materials. The following discussion of the *1S emb* mutants summarizes unpublished results of J.K.Clark. All of the mutants showed cellular and morphological defects, and irregular growth and necrosis occurred in the embryos of all of them. Twinning, predominantly by supernumerary embryo formation from the suspensor or coleorhiza, was observed in four mutants. A

Table 6.3. *The* 1S emb *mutants*

Mutant	No. embryos examined	Allele	Source	Independent mutations
emb1	358	*emb**-8531	*Mutator*	First event
		emb*-8516	*Mutator*	Second event
		emb*-8901	*Ds*	
emb4	318	*emb**-8502	*Mutator*	First event
		emb*-8503	*Mutator*	= 8502
		emb*-8901	*Mutator*	= 8502
emb7	455	*emb**-8519	*Mutator*	First event
		emb*-8520	*Mutator*	= 8519
		emb*-8902	*Ds*	
emb10	468	*emb**-8545	*Mutator*	First event
		emb*-8550	*Mutator*	Second event
		emb*-9402	*Mutator*	Third event
emb12	79	*emb**-8517	*Mutator*	
emb13	130	*emb**-8549	*Mutator*	

worm-like extension of the scutellum was observed in four mutants, and was presumably the result of a local hormone imbalance. These unusual phenotypes suggest that, in the absence of proper functioning of these *emb* genes, the developmental programmes of embryogenesis are disrupted, and new programmes of morphogenesis are either stimulated or de-repressed. Aleurone proliferation occurred in all of the *1S emb* mutants, indicating that these genes function within the endosperm, but in a more tissue-specific manner than the *dek* genes.

The *1S emb* genes are blocked throughout development and have differing phenotypes, indicating that they are involved in diverse developmental processes. However, they fall into the two functional groups identified in the initial examination of the 51 *emb* mutants. Four of the *1S emb* mutants undergo a breakdown in organization. They are characterized by the fact that they are blocked over a relatively broad range of stages, by proliferative growth, and by twinning. Embryos of *emb1* and *emb13* were blocked primarily between the polarization and morphogenesis periods, with death of the embryo proper and proliferation of the suspensor. Embryos of *emb7* and *emb12* were blocked primarily during the morphogenesis period and early maturation periods. These embryos underwent an extremely severe watery necrosis accompanied by cell proliferation. This phenotype is similar to that of the embryo-lethal alleles of the necrotic *dek* mutant *brn1*. The embryos of *emb12* appeared to contain a failed embryonic axis and an inverted shoot apical meristem similar to that of *dek23*.

Two of the *1S emb* mutants show impairment in patterning. They are characterized by being relatively stage-specific in their blocks, and by being uniform in phenotype. Embryos of *emb4* were blocked during the morphogenesis period but lacked an embryonic axis. They were elongate to tubular and wavy in form, and the cells comprising the embryo were abnormally elongated. Some mutant embryos underwent precocious vascular differentiation. Comparison of the phenotype of hypoploid and homozygous diploid mutant embryos indicated that mutation of the *emb4* gene may have disabled a negative regulator involved in shape determination. Embryos of *emb10* showed a variety of pattern defects at the cell, embryo and kernel level. The upper scutellum in most mutant embryos was truncated, and in many cases the embryonic axis was misaligned. Some embryos were wrongly oriented in the kernel, including one noteworthy individual that was upside down and facing backwards. The cells of the upper scutellum and mesocotyl failed to elongate. Zygotic lethality was frequent in the *emb*-9402* allele. It is likely that the *emb10* gene product is required during the first period of embryogenesis in order to ensure correct embryo polarity, during the morphogenesis period to ensure correct siting of the shoot apical meristem, and during the maturation period to promote cell elongation in the upper scutellum and embryonic axis.

Specific aspects of the phenotypes of the *1S emb* mutants are associated with hormone-controlled events. Early embryonic blocks, cell elongation and vascular differentiation are auxin-associated events, whereas mesocotyl elongation in maize is mediated by gibberellic acid. Thus it is likely that the *emb1*, *emb4* and *emb10* genes are involved in hormone signalling pathways. Furthermore, since the general processes of proliferative growth and necrosis are associated with altered auxin levels in genetic and pathogen-induced tumours, it is reasonable to speculate that *emb7* and *emb12* may also be involved in hormone-controlled events.

From their phenotypic diversity it is clear that the genes defined by the *emb* mutations are involved in a wide range of developmental processes. Because of their relative embryo specificity, their early expression and their impact on morphogenesis, it is likely that genes of this class play crucial roles in orchestrating the developmental programmes of maize embryogenesis.

6.5 Future prospects

In considering the phenotypes of the *dek* and *emb* mutations the following conclusions may be drawn.

(i) Many genes are involved in embryo morphogenesis.
(ii) Most are expressed early in embryogenesis.
(iii) Some are involved in the differentiation of specific structures, some with the maintenance of morphological organization and some with patterning processes.

(iv) Many genes that are involved in embryo morphogenesis also mediate morphogenesis of the aleurone.

(v) The pleiotropy of the *dek* and *emb* genes includes not only expression in the embryo and endosperm, but also expression in the gametophyte generation.

Analysis of maize embryogenesis mutants has already identified important developmental genes. In the future it will contribute to the development of more precise models of the genetic programmes underlying embryo morphogenesis, and it will provide valuable tools for evolutionary inquiry.

A genome-based approach offers the opportunity to explore genome structure and evolution. For example, *1S* is the fourth longest of the maize chromosome arms (Sheridan, 1982b). On it reside a number of developmental genes and a quantitative trait locus (QTL) conditioning embryogenic response in culture (Armstrong *et al.*, 1992; Coe *et al.*, 1988). Much of *1S* is duplicated on *9L* (M. Murray, personal communication, *Maize Genet. Coop. Newslett.* **62**, 89). Mapping of the *1S emb* genes will enable the functional significance of their genomic context to be examined. Once they are cloned they can be used as probes to isolate, map and characterize their duplicated counterparts within the maize genome. On a larger scale, comparative linkage analysis, combined with our rapidly expanding knowledge of duplicated regions in the maize genome and the collinearity of all cereal genomes, will provide insight into the evolution of the grass genome (Bennetzen and Freeling, 1993; Shields, 1993). Comparative mapping has shown that the duplicated maize arms *1S* and *9L* are homologous and partially orthologous with rice chromosome *3*, which is in turn homologous with the long arm of wheat chromosome *4* (Ahn and Tanksley, 1993; Ahn *et al.*, 1993). Therefore maize, wheat and rice represent distant points in a long sequence of divergence and duplication, and arm-by-arm comparisons of the *dek* and *emb* genes may reveal developmentally significant genome structure that has been preserved over time.

Comparative structural and expression analysis of the genes defined by maize embryogenesis mutants can be used to study the evolution of the grasses. The Gramineae comprises only a few major groups at the tribe and subfamily level (Kellogg and Campbell, 1987). The grass tribes can be distinguished strictly on the basis of embryo morphology (Reeder, 1957; Takeota, 1964). These groupings are consistent with phylogenetic trees based on molecular and morphological characters (Kellogg and Watson, 1993). This indicates that, in the grasses, the developmental programmes of embryogenesis are correlated with the fundamental morphological changes which have divided major evolutionary groups. The maize embryogenesis mutants may include genes that are involved in determining key taxonomic traits. For example, the extent of mesocotyl development is a key trait distinguishing the panicoid grasses (Reeder, 1957). Maize and wheat differ in the degree of mesocotyl cell elongation (Avery, 1930), and changes in the structure and expression of the *emb10* gene may have played a role in the evolution of this morphological change. It has been predicted that 'increasing integration of genetics and phylogenetics will undoubtedly lead to new concepts of the nature of the evolutionary process and to new hypotheses of mechanisms' (Kellogg and Birchler, 1993). The maize *dek* and *emb* genes offer an excellent opportunity to identify significant changes in the embryogenesis genes of maize, wheat, barley and rice, and to examine these changes in the light of evolution.

Cloning of the maize embryogenesis genes will provide tools for studying the evolution of the moncot embryo. Two longstanding questions concern the initial position of the monocot apical meristem and the derivation of the regions of the embryo. An *emb10* clone may provide a marker to determine whether the shoot apical meristem is initiated terminally or laterally in normal embryos (Guignard, 1984). The *emb10* gene may also be used to investigate whether the embryo has evolved as a composite structure. An *emb10* clone may provide a marker for two evolutionary and ontogenetic compartments of the grass cotyledon, namely the leaf-like suctorial organ (upper scutellum) and the neck (mesocotyl; reviewed by Eames, 1961). Similarly, *fl*-1253B* and other mutants which undergo proliferation of the coleorhiza and lower scutellum may provide markers for a putative embryonic compartment homologous to the 'foot' of primitive taxa (Vallade *et al.*, 1993).

The genetic resources of maize are readily accessible. Protocols for handling maize as an experimental organism, for genetic manipulations, and for analysis of mutants may be found in Sheridan (1982a) and Walbot and Freeling (1993). An overview of maize genetics has been provided by Coe *et al.* (1988) and Neuffer *et al.* (in press). Seed of maize mutants and genetic stocks may be obtained from the Maize Stock Center (University of Illinois) and the University of Missouri Maize RFLP Laboratory distributes DNA probes. Informal exchange of information is provided by the Maize Genetics Cooperation Newsletter. Extensive information is available on-line from the Maize Genome Database at http://www.agron. missouri.edu./top.html.

Maize embryogenesis mutants have constituted an enormous resource for understanding embryo development in maize. They will become even more valuable in the future as old mutants are analysed in more detail, new mutants are recovered and studied, and information about the maize embryogenesis genes is correlated with that obtained for other grasses and other plant species.

References

Abbé, E.C. and Stein, O.L. (1954) The growth of the shoot apex in maize: embryogeny. *Am. J. Bot.* **41**, 285–293.

Ahn, S. and Tanksley, S.D. (1993) Comparative linkage maps of the rice and maize genomes. *Proc. Natl Acad. Sci. USA* **90**, 7980–7984.

Ahn, S., Anderson, J.A., Sorrells, M.E. and Tanksley, S.D. (1993) Homologous relationships of rice, wheat and maize chromosomes. *Mol. Gen. Genet.* **241**, 483–490.

Armstrong, C.L., Romero-Severson, J. and Hodges, T.K. (1992) Improved tissue culture response to an elite maize inbred through backcross breeding, and identification of chromosomal regions important for regeneration by RFLP analysis. *Theor. Appl. Genet.* **84**, 755–762.

Avery, G.S. Jr (1930) Comparative anatomy and morphology of embryos and seedlings of maize, oats, and wheat. *Bot. Gaz.* **89**, 1-39.

Bennetzen, J.L. and Freeling, M. (1993) Grasses as a single genetic system: genome composition, collinearity and compatibility. *Trends Genet.* **9**, 259-260.

Brink, R.A. (1927) A lethal mutation in maize affecting the seed. *Am. Naturalist* **61**, 520-530.

Clark, J.K. and Sheridan, W.F. (1986) Developmental profiles of the maize embryo-lethal mutants *dek22* and *dek23*. *J. Hered.* **77**, 83-92.

Clark, J.K. and Sheridan, W.F. (1988) Characterization of the two maize embryo-lethal defective kernel mutants *rgh*-1210* and *fl*-1253B*: effects on embryo and gametophyte development. *Genetics* **120**, 279-290.

Clark, J.K. and Sheridan, W.F. (1991) Isolation and characterization of 51 embryo-specific mutations of maize. *Plant Cell* **3**, 935-951.

Coe, E.H. Jr, Neuffer, M.G. and Hoisington, D.A. (1988) The genetics of corn. In: *Corn and Corn Improvement*, 3rd edn (eds G.S. Sprague and J.W. Dudley). American Society of Agronomy, Crop Science Society of America, Soil Science Society of America, Madison, pp. 81-258.

Coulter, M.C. (1925) A distortion of the 3:1 ratio. *Bot. Gaz.* **79**, 28-44.

Demerec, M. (1923) Heritable characters of maize XV. Germless seed. *J. Hered.* **114**, 277-300.

Diboll, A.G. (1968) Fine structural development of the megagametophyte of *Zea mays* following fertilization. *Am. J. Bot.* **55**, 797-806.

Diboll, A.G. and Larson, D.A. (1966) An electron microscope study of the mature megagametophyte in *Zea mays*. *Am. J. Bot.* **53**, 391-402.

Eames, A.J. (1961) *Morphology of the Angiosperms*. McGraw-Hill, New York.

Emerson, R.A. (1932) A recessive zygotic lethal resulting in 2:1 ratios for normal vs. male-sterile and colored vs. colorless pericarp in F_2 of certain maize hybrids. *Science* **75**, 566.

Emerson, R.A. (1939) A zygotic lethal in chromosome 1 of maize and its linkage with neighboring genes. *Genetics* **24**, 368-384.

Gavazzi, G., Nava-Rachi, M. and Tonelli, C. (1975) A mutation causing proline requirement in maize. *J. Theor. Appl. Genet.* **46**, 339-346.

Goodsell, S.M. (1927) *The Relation Between Numbers of Defects Present and Yield of Corn.* Ph.D. Thesis, Iowa State College.

Guignard, J.-L. (1984) The development of cotyledon and shoot apex in monocotyledons. *Can. J. Bot.* **62**, 1316-1318.

Horovitz, S. and Sanguineti, M.E. (1936) *Siamensis*, nuevo caracter hereditario del maiz. *Rev. Argent. Agron.* **3**, 245-249.

Johri, B.M. (ed.). (1984) *Embryology of the Angiosperms*. Springer, Berlin.

Johri, M.M. and Coe, E.H. (1982) Genetic approaches to meristem organization. In: *Maize for Biological Research* (ed. W.F. Sheridan). Plant Molecular Biology Association, Charlottesville, pp. 301-310.

Jones, D.F. (1920) Heritable characters of maize. IV. A lethal factor — defective seeds. *J. Hered.* **11**, 161-167.

Kellogg, E. A. and Birchler, J.A. (1993) Linking phylogeny and genetics: *Zea mays* as a tool for phylogenetic studies. *Syst. Bot.* **42**, 415-439.

Kellogg, E.A. and Campbell, C.S. (1987) Phylogenetic analyses of the Gramineae. In: *Grass Systematics and Evolution* (eds T.R. Soderstrom, K.W. Hilu, C.S. Campbell and M.E. Barkworth). Smithsonian Institution Press, Washington, pp. 310-322.

Kellogg, E.A. and L. Watson, L. (1993). Phylogenetic studies of a large data set. I. Bambusoideae, Andropogonodae, and Pooideae (Gramineae). *Bot. Rev.* **59**, 273-343.

Kiesselbach T.A. (1949) The Structure and Reproduction of Corn. University of Nebraska Press, Lincoln.

Kiesselbach, T.A. and Walker, E.R. (1952) Structure of certain specialized tissues in the kernel of corn. *Am. J. Botany* **39**, 561-569.

Kyle, D.J. and Styles E.D. (1977) Aleurone development in maize. *Planta* **137**, 185-193.

Maheshwari, P. (1950) An Introduction to the Embryology of the Angiosperms. McGraw-Hill, New York.

Manglesdorf, P.C. (1923) The ingeritance of defective seeds in maize. *J. Hered.* **14**, 119-125.

Manglesdorf, P.C. (1926) The genetics and morphology of some endosperm characters in maize. *Conn. Agric. Stn. Bull.* **279**, 509-614.

Neuffer, M.G. and Sheridan, W.F. (1980). Defective kernel mutants of maize. I. Genetic and lethality studies. *Genetics* **95**, 929-944.

Neuffer, M.G., Chang, M.T., Clark, J.K. and Sheridan, W.F. (1986) The genetic control of maize kernel development. In: *Regulation of Carbon and Nitrogen Reduction and Utilization in Maize* (eds. C. Shannon, P. Knievel and C.D. Boyer). American Society of Plant Physiologists, Rockville, Maryland, pp. 35-50.

Neuffer, M.G., Coe, E.H. and Wessler, S. (in press) *The Mutants of Maize*, Cold Spring Harbor Press, Cold Spring Harbor.

Paxson, J.B. (1963) *Some effects of plant regulators on embryogeny in Zea mays.* Ph.D. Thesis, Texas A. and M. University, College Station.

Poethig, R.S., Coe, E.H. Jr and Johri, M.M. (1986) Cell lineage patterns in maize embryogenesis: a clonal analysis. *Dev. Biol.* **117**, 392-404.

Raghavan, V. (1986) *Embryogenesis in Angiosperms*. Cambridge University Press, Cambridge.

Randolph, L.F. (1936) Developmental morphology of the caryopsis in maize. *J. Agric. Res.* **53**, 881-916.

Reeder, J.R. (1957) The embryo in grass systematics. *Am. J. Bot.* **44**, 756-768.

Rost, T.L. and Lersten, N.R. (1973) A synopsis and selected bibliography of grass caryopsis anatomy and fine structure. *Iowa State J. Res.* **48**, 47-87.

Sass J.E. (1977) Morphology. In: *Corn and Corn Improvement*, 2nd Edn. (ed. G.S. Sprague). American Society of Agronomy, Crop Science Society of America, Soil Science Society of America, Madison, pp. 89-110.

Sass, J.E. and Sprague, G.F. (1949) Histological development of 'Accessory Blade' and associated abnormalities in maize. *Iowa State Coll. J. Sci.* **23**, 301-309.

Sass, J.E. and Sprague, G.F. (1950) The embryology of 'germless' maize. *Iowa State Coll. J. Sci.* **24**, 209-218.

Scanlon, M.J., Stinard, P.S., James, M.G., Myers, A.M. and Robertson, D.S. (1994) Genetic analysis of sixty-three mutations affecting maize kernel development isolated from *Mutator* stocks. *Genetics* **136**, 281-294.

Schwartz, B.W., Yeung, E.C. and Meinke, D.W. (1994) Disruption of morphogenesis and transformation of the suspensor in abnormal suspensor mutants of *Arabidopsis*. *Development* **120**, 3235-3245.

Sheridan, W.F. (1982a) *Maize for Biological Research*. Plant Molecular Biology Association, Charlottesviile.

Sheridan, W.F. (1982b) Maps, markers, and stocks. In: *Maize for Biological Research* (ed. W.F. Sheridan). Plant Molecular Biology Association, Charlottesville, pp. 37-52.

Sheridan, W.F. and Clark, J.K. (1987) Maize embryogeny, a promising experimental system. *Trends Genet.* **3**, 3–6.

Sheridan, W.F. and Clark, J.K. (1993a) Fertilization and embryogeny in maize. In: *The Maize Handbook* (eds M. Freeling and V. Walbot). Springer-Verlag, New York, pp. 647–652.

Sheridan, W.F. and Clark, J.K. (1993b) Mutational analysis of morphogenesis of the maize embryo. *Plant J.* **3**, 347–358.

Sheridan, W.F. and Neuffer, M.G. (1980) Defective kernel mutants of maize. II. Morphological and embryo culture studies. *Genetics* **95**, 945–960.

Sheridan, W.F. and Neuffer, M.G. (1981) Maize mutants altered in embryo development. In: *Levels of Genetic Control in Development* (eds S. Subtelney and U. Abbott). Alan R. Liss, New York, pp. 137–156.

Sheridan, W.F. and Neuffer, M.G. (1982) Maize developmental mutants: embryos unable to form leaf primordia. *J. Hered.* **73**, 318–329.

Sheridan, W.F. and Neuffer, M.G. (1986) Genetic control of embryo and endosperm development in maize. In: *Gene Structure and Function in Higher Plants* (eds G.M. Reddy and E.H. Coe, Jr). Oxford and IBH Publishing Co., New Delhi, pp. 105–122.

Sheridan, W.F. and Thorstenson, Y.R. (1986) Developmental profiles of three embryo-lethal maize mutants lacking leaf primordia, *ptd*-1130, cp*-1418*, and *bno*-747B. Dev. Genet.* **7**, 35–49.

Shields, R. (1993) Pastoral synteny. *Nature* **365**, 297–298.

Sprague, G.F. (1941) Transmission tests of maize mutants induced by ultra-violet radiation. *Iowa Agric. Exp. Stn Res. Bull.* **292**, 389–407.

Takeota, T. (1964) Notes on some grasses. XVI. Embryo structure of the genus *Oryza* in relation to the systematics. *Am. J. Bot.* **51**, 539–543.

Vallade, J., Bugnon, F. and Ibannain, S. (1993) Interpretation morphologique de l'embryon chez les Embryophytes, avec application au cas des Graminees (Poaceae). *Can. J. Bot.* **71**, 256–272.

VanLammeren, A.A. (1986) Developmental morphology and cytology of the young maize embryo (*Zea mays* L.) *Acta Bot. Neerl.* **35**, 169–188.

VanLammeren, A.A. (1987) *Embryogenesis in* Zea mays L. *A Structural Approach to Maize Caryopsis Development* in Vivo *and* in Vitro. Ph.D. Thesis, University of Wageningen.

Walbot, V. and Freeling, M. (eds) (1993) *The Maize Handbook*. Springer-Verlag, New York.

Wang, F.H. (1947) Embryological development of inbred and hybrid *Zea mays. Am. J. Bot.* **34**, 113–125.

Wentz, J.B. (1930) The inheritance of germless seeds in maize. *Iowa Exp. Stn Res. Bull.* **121**, 347–379.

Wolf, M.J., Buzan, C.L., MacMasters, M.M. and Rist, C.R. (1952) Structure of the mature corn kernel. IV. Microscopic structure of the germ of dent corn. *Cereal Chem.* **29**, 362–382.

Gene expression and embryonic maturation in cereals

A.C. Cuming, M. Türet and W. Butler

7.1 Introduction

Development in cereals is initiated by fertilization, an event which gives rise to a small diploid embryo and a massive triploid endosperm. Despite the relative disparity in size of the embryo and endosperm, the cereal embryo at maturity appears to be a more highly differentiated structure than that of dicotyledonous species, in which the cotyledons rather than the endosperm generally serve as the principal repositories of storage reserves. Whereas the dicotyledonous embryo, exemplified in Chapters 2, 3 and 7 of this volume by the embryo of *Arabidopsis thaliana*, comprises merely the root and shoot apical meristems, and the cotyledons as recognizable organs (although with the fundamentals of seedling pattern already established), cereal embryos typically possess a more complex anatomy, consisting of both primary and lateral root primordia and a shoot apical meristem which, at maturity, has generated a significant number of enveloping leaf primordia. These structures possess differentiated vascular connections to each other and to the single cotyledon (the scutellum, modified for the storage of embryonic reserves and the transmission of endosperm reserves to the embryonic axis). The embryonic axis is additionally sheathed within enveloping organs, whose existence can be considered as transient in the development of the plant; the coleorhiza surrounds the root and the coleoptile surrounds the shoot. The mature embryo contains, in miniature, all of the structures recognizable in the vegetatively growing plant. These are merely elaborated following germination. The process of embryogenesis thus consists of all the morphogenetic events that occur within the plant until the initiation of flowering causes the terminal transition of the shoot apex from a progenitor of purely vegetative structures, such that it becomes reprogrammed to generate the reproductive organs.

The temporal development of the cereal embryo can be divided into a succession of distinct stages. Immediately following fertilization, cellular proliferation and differentiation take place, and are accompanied by establishment of the characteristic embryonic pattern. Until recently, progress towards a fuller understanding of the molecular processes underlying pattern formation had been hindered by the relative intractability of the cereal embryo to biochemical procedures. However, advances in the use of molecular genetic techniques, particularly transposon-mediated mutagenesis and subsequent gene tagging, in maize, are opening up new avenues of research. The application of this approach is described in Chapter 6 by J. Clark. Tissue culture, leading to the generation of somatic embryos, is also becoming increasingly important as an alternative means of analysing the regulatory events in the formation of a differentiated embryo. Although the greatest advances have been made in the production of somatic embryos from the more amenable dicotyledonous species (as described by Toonen and de Vries, Chapter 11), an understanding of the underlying molecular processes in the generation of somatic embryos from cereal cell cultures is likely to be of increasing importance, as this approach offers the most promising route to the production of transgenic cereals.

In this chapter, we shall mainly address the events which occur subsequent to the formation of the differentiated embryo, but prior to the final maturation of the embryo. Although not concerned with morphogenesis *per se*, these later events are still intrinsic to embryogenesis. During this period, a developmental choice is made which commits the embryo to a period of apparent arrest, and this stage is typified by the accumulation of embryonic reserves and the acquisition of tolerance to environmental insults. This period is essential for the formation of a mature and robust seed, capable of undergoing storage for prolonged periods in a dry state with little loss of viability. This property has been a major factor, perhaps the principal one, in the success of the cereals in cultivation. The ability to store surplus seed from one year to the next was a formative influence in the development of the agricultural economies which have represented the foundation of human civilizations.

7.2 The wheat embryo as an experimental model

Our studies have been concerned largely with the molecular biology of embryogenesis in wheat (*Triticum aestivum* L.), but many of the conclusions we have drawn are equally applicable to other cereal species. In wheat, embryogenesis can be viewed as progressing through successive stages, which can be characterized by the processes which occur within the embryo and by the increasing age of the embryo. Because the rate of development of the embryo is dependent on the prevailing environmental conditions, particularly temperature and humidity (Bennett *et al.*, 1973), it is more convenient to use a system of physiological development than one based on temporal age. Such systems are based on the morphological characteristics of the developing embryo, and throughout this chapter we shall refer to the system introduced by Rogers and Quatrano (1983), which recognizes five stages of development. Stages 1 and stage 2 represent the earliest periods of development, during which cell proliferation and differentiation establish the characteristic embryonic morphology. Under the conditions used in our laboratory (plants grown in a 16-h light, 25°C/8-h dark, 18°C cycle at 70% relative humidity (RH)) these stages are complete at 10–12 days post-anthesis (dpa). Only after the completion of stage 2 have the immature embryos acquired the ability to germinate in culture, when dissected from the developing grain (Morris *et al.*, 1990). The embryo embarking on stage 3 of its development stands at a crossroads in terms of its commitment to a developmental fate. It has achieved 'functional maturity': the ability to proceed immediately, by precocious germination, to establish a seedling and to embark on vegetative growth. In the course of normal development, however, the embryo does not follow such a course. Indeed, to germinate at this early stage in the development of the grain would be detrimental to the survival of the seedling. Because the development of its sister tissue, the endosperm, follows a temporally less accelerated course (cf. Brown *et al.*, Chapter 10), the accumulation of reserve substances within the endosperm at this time is far from complete. A precociously germinating seedling would thus risk exhausting the available reserves before becoming sufficiently established to support itself autotrophically. Thousands of years of human selection have sought to eliminate such a possibility. Instead, the mature wheat embryo assumes a state of developmental arrest. It remains in a dormant condition, within the developing seed, until grain filling is complete. Usually this period is terminated by a progressive dehydration of the ear — a part of the programmed senescence of the parent plant — and following harvest the grains may be further dehydrated to moisture levels as low as *c.* 5% for prolonged storage without incurring injury.

The dormant condition is not one of metabolic inactivity. The embryo can be observed to increase in both size and content. During this period the embryonic reserves — storage proteins and lipid bodies — accumulate, as do a number of additional products which have been implicated in the acquisition of desiccation tolerance. The accumulation of these compounds (principally disaccharides and a subset of polypeptides) occurs throughout stages 3 and 4, culminating in the production of the mature, dry embryo (stage 5) (Quatrano *et al.*, 1986). During stages 3 and 4, there is progressive water loss from the grain. The desiccation of the grain must be viewed as an inherent part of the embryogenic programme. It acts as an environmental trigger which switches the developmental commitment of the embryo from a pathway characterized by developmental arrest and anabolic metabolism in stages 3 and 4 to one of catabolic metabolism of the stored reserves to fuel the rapid germinative growth which commences immediately after the imbibition of water by the dry grain.

An indication of the developmental flexibility of the embryo at stage 3 of its development is gained from observing the manner in which its fate may be determined by the application of growth regulators. When immature embryos are dissected from immature grains at the beginning of stage 3, and are incubated on a culture medium containing only simple nutrients, they will undergo precocious germination. This can be prevented in two ways. If the medium is supplemented with the plant growth regulator abscisic acid (ABA), the embryos do not germinate. Instead they remain in an arrested state, and undergo molecular changes characteristic of embryonic dormancy (see below) (Quatrano *et al.*, 1986). Alternatively, the developmental potential of the embryo can be reversed; instead of the cells becoming irrevocably committed to the formation of the differentiated organs of the mature embryo, treatment with the synthetic auxin 2,4-D initiates the proliferation of dedifferentiated cells from within the embryo, resulting in the formation of a mass of callus tissue. However, such callus tissue retains its embryogenic potential. Subsequent withdrawal of the auxin can result in the regeneration of differentiated plants through a process of somatic embryogenesis (Maddock and Lancaster, 1983), and the dedifferentiated callus tissue retains the ability to express genes characteristic of embryonic maturation, if stimulated appropriately (see below). The developmental pathways open to the embryo are illustrated as a flow chart in Figure 7.1, and the morphological changes which take place during embryonic maturation, germination and callus formation are shown in Figure 7.2.

7.3 Molecular markers of embryo development

What causes the embryo to forgo the opportunity to germinate during developmental stage 3, yet to embrace this option, with alacrity, after stage 5? We have analysed the changes in gene expression which accompany this change in

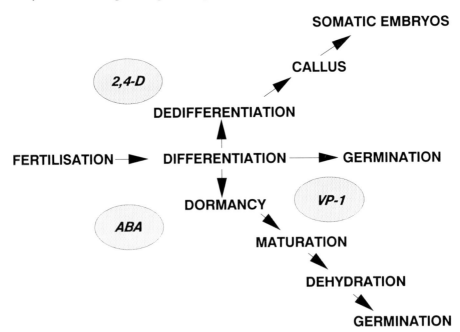

Figure 7.1. *Developmental fates open to the wheat embryo.*

developmental state in order to identify the molecular markers which indicate the underlying developmental switch. Such markers were initially identified by examining the changes in gene expression which occurred when dry embryos (stage 5) underwent germination in the presence of water (Cuming, 1984). The gene products (mRNA and proteins) present in the dry embryo represent the products of transcriptional activity during the preceding stages of development, namely the genes expressed during stages 3 and 4. A simple experiment, which compared the pattern of proteins present in the dry embryo with that of proteins synthesized during the initial hours of imbibition, revealed a number of significant changes in the population of embryonic polypeptides. Most prominent of these was the disappearance of two highly abundant polypeptides: one of apparent M_r *c*. 14 000, and another of apparent M_r *c*. 50 000. The former has been designated by the acronym 'Em polypeptide', originally for '*E*arly, *M*ethionine-labelled polypeptide' (Grzelczak *et al.*, 1982). This refers to the observation that, following labelling *in vivo* with [^{35}S]methionine, the protein was labelled early but not at a late stage of imbibition. In retrospect, this acronym has proved to be somewhat misleading, in that the protein is not particularly methionine-rich, and because it is sometimes mistakenly assumed to be associated with germination rather than with embryogenesis. For this reason, we favour the use of the designation '*E*mbryo *M*aturation polypeptide' to denote, more accurately, its developmental specificity. The larger polypeptide has since been identified as the principal storage globulin of the embryo (Türet, 1993). Homologues of both the Em polypeptide and the storage globulin have been identified in other higher plants, including both monocotyledonous and dicotyledonous species, where their patterns of expression have been shown to be similar to that observed during wheat embryo maturation. By contrast with these gene products, a third molecular marker has been identified whose expression is typical of germination (Grzelczak and Lane, 1984). Originally named 'germin', this is an extracellular, homopentameric glycoprotein which has now been shown to be an oxalate oxidase (Dumas *et al.*, 1993; Lane *et al.*, 1993). The germin monomer is a polypeptide of M_r *c*. 25 000.

Cloned sequences encoding these proteins have been isolated and used to examine their temporal patterns of expression during development. Northern blot analysis of mRNA from developing embryos has shown that the two embryogenesis-related genes (or gene families, since each is represented by a small multigene family in each of the three contributory genomes of hexaploid wheat) have overlapping but distinct patterns of expression (Figure 7.3a and b). The expression of both genes is first detected as the embryo enters developmental stage 3. However, whereas a high level of globulin mRNA accumulates rapidly at stage 3, and thereafter exhibits a slow decline throughout subsequent development (becoming almost undetectable in the mature, stage 5, embryo), the mRNA encoding the 'Em' polypeptide starts to accumulate somewhat later, and its levels increase throughout stages 3 and 4. The Em mRNA remains present in the dry embryo, comprising perhaps the single most abundant species in this messenger population, until the uptake of water when it disappears within 12 h of the onset of imbibition. The latter pattern of expression is typical of many embryonic

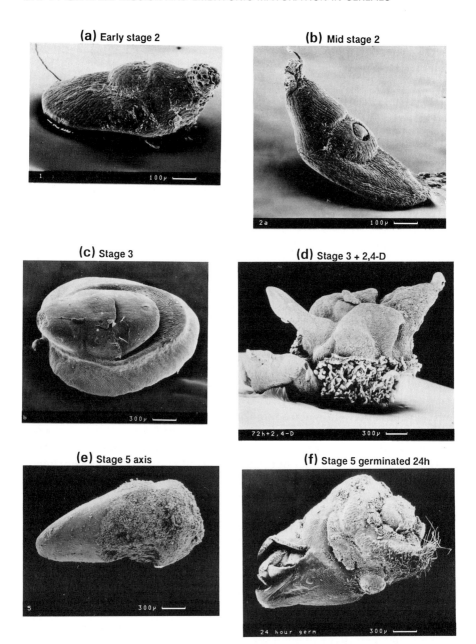

Figure 7.2. *Characteristics of wheat embryo development. (a) An embryo early in stage 2 of development — the embryonic axis, sheathed within the coleoptile and coleorhiza, has become distinct from the underlying scutellum. Note the highly structured cells at the tip of the coleorhiza, which represent the residue of the suspensor. (b) An embryo later in stage 2. The coleoptile has not completed its growth, and a leaf primordium can be seen through the aperture. (c) An embryo in stage 3. The embryonic axis and scutellum have undergone*

non-storage proteins identified in both dicotyledonous and monocotyledonous plants. This has led to these proteins and their messengers being generically described as 'Late Embryogenesis Abundant' or *LEA* gene products (Galau *et al.*, 1986). Germin mRNA is not present in immature or dry embryos, but only accumulates following imbibition of the dry embryo, or upon the precocious germination of isolated immature embryos in culture (Figure 7.3d). Like the embryogenesis-related proteins, germin is also the product of a member of a multigene family in wheat (Lane *et al.*, 1991). A number of germin-like proteins have been identified in a range of other plant species, but the full extent of this gene family and the patterns of expression of its members have yet to be characterized as extensively as those of the Em and globulin gene families.

7.4 Abscisic acid and osmotic stress as regulatory agents

A feature of the expression of the two maturation-related gene products is that both may be induced precociously by the application of the growth regulator ABA. This growth regulator has long been associated with the control of seed dormancy, and there is compelling evidence that it controls the switch between dormancy and maturation, and precocious germination, that operates at the interface between developmental stages 2 and 3. First, it is generally believed that a close correlation exists between the ABA concentration and the developmental stage of the seed, with high levels of ABA being present at the time of the onset of dormancy (a view that is not conclusively supported by an extensive literature; for a recent critique, see Hilhorst, 1995). However, as noted above, although immature (stage 3) embryos will germinate precociously when removed from the confines of the developing grain and placed on nutrient medium, they will not do so if this medium is supplemented with ABA. Finally,

---

considerable enlargement, and the coleoptile has completely enveloped the shoot primordia, leaving only a narrow slit through which the leaf primordia will emerge on germination. (d) An embryo harvested at the same stage as that in (c), which has been incubated in the presence of 2,4-D for 72 h of development. Growth of the root and shoot, typical of germination, has been suppressed, and cell proliferation within the embryo has led to its gross distortion, followed by splitting of the junction between the embryonic axis and the scutellum. The characteristically spiral callus cells are spilling out of the rift. (e) An embryonic axis obtained from a dry (stage 5) embryo. This view from the ventral side shows a region of damaged tissue where the axis was sheared from the scutellum. The coleoptile extends to the left of the panel. (f) An embryonic axis at stage 5, which has imbibed water for 24 hours. This view from the dorsal surface shows that germination has been initiated. The primary root can be seen emerging from the right-hand end of the axis, with a lateral root emerging below. The coleoptile has split and the elongating leaf primordia can be seen emerging at the left-hand end.

(a) Globulin mRNA

								hpi	14 dpa + 3d incubation on

dpa
14 15 16 18 19 22 26 31 48 MS 10^{-7} 10^{-6} 10^{-5}M ABA

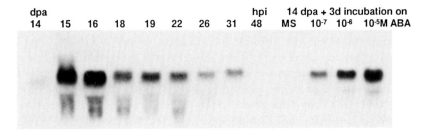

(b) Em mRNA

dpa   Mature embryos
14 15 16 17 18 19 20   Dry 40' 2h 6h imbibition

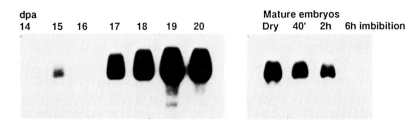

(c) Em mRNA

dpa
14 20 ABA 5% 10% 15% 20% mannitol

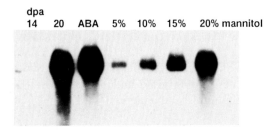

(d) Germin mRNA

Mature embryos   12dpa embryos precocious germination
Dry 24h 48h imbibed   3h   6h   16h 24h

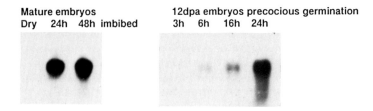

Figure 7.3. *Expression of mRNA markers of wheat development. Northern blot analysis of globulin, Em and germin mRNA during wheat embryo development; total RNA (10 μg) from embryos treated as indicated was resolved by agarose gel electrophoresis prior to hybridization with labelled cDNA probes. (a) Globulin mRNA. RNA was isolated from embryos at successive stages during embryogenesis (14–31 dpa), from mature, dry embryos imbibed for 48 h (48 hpi) and from immature (14 dpa) embryos cultured on Murashige.*

in another cereal, *Zea mays* L., there are a number of well-characterized mutants which do not exhibit a period of dormancy. These so-called 'viviparous' (*vp*) mutants undergo precocious germination while the immature kernels are still within the developing grain (Robertson, 1955). In many of these mutants (*vp-2–vp-9*) the metabolic pathway leading to the synthesis of ABA is blocked (Neill *et al.*, 1986, 1987), whilst another mutant, *vp-1*, has an ABA-insensitive phenotype. The isolation of additional viviparous mutants is described by McCarty in Chapter 8 of this volume.

The availability of water is also essential for germination of the embryo, and as the developing embryo becomes older, the water content of the immature grain becomes progressively reduced. This probably also contributes to the dormancy of the embryo during the period of maturation, and this factor can be used as an experimental tool. Precocious germination of isolated immature embryos can be prevented if their culture medium is supplemented with an osmotically active agent. The growth of plant cells is driven by turgor pressure resulting from net water uptake by cells, and incubation on a medium of high osmotic strength prevents such uptake and thereby physically blocks cell expansion. This phenomenon is illustrated by the effects of supplementing the culture medium with increasing concentrations of mannitol, a non-ionic osmolyte. Under these conditions the precocious germination of immature embryos is suppressed, and their rate of growth is reduced in proportion to the degree of stress applied (Morris *et al.*, 1990).

Both ABA treatment and the application of an osmotic stress will induce the expression of maturation-specific genes, and suppress the induction of germination-specific genes. Thus, both globulin mRNA and Em mRNA are induced by these treatments, whereas the expression of germin mRNA only occurs in the absence of these germination-suppressive agents (Figure 7.3). We have found that both mRNA species accumulate in a dose-dependent manner with respect to the inducing stimulus. Figure 7.3a shows the ABA concentration dependence of gene expression for the globulin mRNA, while the response of the Em genes to osmoticum is seen to increase under progressively more stressful conditions (Figure 7.3c). Such inducibility by osmotic stress and ABA is also a characteristic of other *LEA* gene products, and a brief examination of the literature will reveal a number of different designations based on these properties. Perhaps the most commonly used are '*RAB*' genes (for **R**esponsive to **AB**A: Mundy and Chua, 1988)

Skoog (MS) medium, supplemented with ABA (10^{-7}–10^{-5}M). (b) Em mRNA. RNA was isolated from immature embryos at successive stages of embryogenesis (14–20 dpa) and from mature, dry embryos imbibed with water for 40 min, 2 h and 6 h. (c) Em mRNA. RNA was isolated from embryos at 14 dpa and 20 dpa, and from 14 dpa embryos subsequently cultured in the presence of 10^{-5}M ABA or increasing concentrations of mannitol (5–20%, w/v). (d) Germin mRNA. RNA was isolated from mature, dry embryos imbibed with water for 24 and 48 h, and from 12 dpa embryos allowed to germinate precociously on MS medium for 3–24 h.

and 'dehydrins', which indicate a specific subset of dehydration-induced '*RAB*' proteins (Close *et al.*, 1989). These common properties of inducibility and temporal pattern of accumulation have led to the suggestion that the function of the non-storage *LEA* proteins may be related to the acquisition of desiccation tolerance. This phenomenon is discussed briefly below, and in more detail by Bartels *et al.* in Chapter 9 of this volume.

The tolerance of desiccation by plants is a property shared by mature seeds, vegetative tissues of some lower plants and vegetative tissues of a very few higher plants. Water is required in the cell not only as a general solvent in which biochemical reactions can proceed, but also as a structural component of macromolecules. Proteins typically have a component of 'bound water' which plays an essential role in the maintenance of their tertiary structure (Edsall and McKenzie, 1983), and cellular membranes maintain their integrity as barriers to permeability through their hydrophobic interactions with the aqueous phases of the cell. The progressive loss of water from the cell necessarily causes an increase in the concentration of solutes, which in the case of ionic solutes could exert toxic effects through the disruption of the normal ionic balance of the cell, and with increasing dehydration would eventually lead to the loss of that fraction of the cell's water which is associated with macromolecular structures. This would result in the irreversible denaturation of such structures, in the case of proteins, and the disruption of membrane structure, as the phospholipid components of the membrane undergo phase changes at high levels of dehydration (McKersie and Senaratna, 1983). The responses of resistant cells and tissues to water loss must overcome these problems. Initially, the accumulation of compatible osmolytes, typically non-ionic solutes, will maintain an osmotic balance between the intracellular and extracellular environments (Rhodes and Hanson, 1993; Yancey *et al.*, 1982). Under extreme conditions of dehydration, however, other mechanisms must exist to protect the cell. Several suggestions have been offered as to the nature of such protective mechanisms. In embryos, these typically invoke a role for those products which are seen to accumulate within the embryo during the period of maturation. The accumulation of disaccharides, sucrose and raffinose being the principal sugars accumulated in cereal embryos, has been linked to a change in the physical state of water in the embryo. Physical measurements indicate that, as the embryo becomes more dehydrated, so the cytosol undergoes a vitreous transformation and becomes glass-like (Leopold *et al.*, 1992). Such a change would result in a substantial decrease in diffusion of components, and would prevent associations between incompatible macromolecules which might lead to the formation of denatured complexes.

In addition, the preservation of membrane properties can be achieved through close association with polyhydric alcohols, which represent a major component of the compatible solutes that accumulate in stressed tissues (Crowe and Crowe, 1992; Vernon and Bohnert, 1992). Several roles have been proposed for the polypeptides which accumulate at this stage of development. Most of the non-storage *LEA* proteins are highly hydrophilic and contain many charged and uncharged but polar amino acid residues. The Em polypeptide is a typical member of this class of protein, and has the additional property of existing as a highly flexible random coil (McCubbin *et al.*, 1985). Proteins with these properties could associate with other macromolecules through hydrogen-bonding interactions, gradually replacing water molecules that have been lost during dehydration. Furthermore, they could sequester the increasing concentration of ions accumulating in the cell, through electrostatic interactions. Such a role has been proposed for one class of *LEA* protein which contains motifs with the capacity to form amphipathic α-helices. These might form complexes via intermolecular hydrophobic associations, presenting charged residues to the exterior for ion binding (Dure, 1993). Alternatively, because many *LEA* proteins have a very high potential for retaining water in bound form, they might act as 'molecular sponges', retaining a shell of hydration and sharing this with other macromolecules in a quasi-chaperone relationship (Lane, 1991). Unfortunately, all of these potential functions are based on correlative, rather than direct evidence. The accumulation of *LEA* proteins occurs at the time when desiccation tolerance is acquired, and the expression of their genes is initiated by stress and stress-related stimuli. However, except for those polypeptides for which an enzymatic function has been demonstrated (Bartels *et al.*, Chapter 9 of this volume; Vernon and Bohnert, 1992), their potential functions remain little more than speculative. Attempts to enhance the desiccation tolerance of otherwise intolerant tissues by the ectopic expression of dehydration-induced genes in transgenic plants have not been successful (Iturriaga *et al.*, 1992), and the elimination of the expression of individual *LEA* genes by antisense RNA has been similarly uninformative (M. Robertson, personal communication). This is perhaps not surprising in view of the complexity of the phenomenon of desiccation tolerance and the broad spectrum of compounds whose accumulation has been associated with its acquisition.

The relationship between ABA and osmotic stress during embryo development is not a simple one. It has long been established that ABA plays a central role in co-ordinating plant responses to water stress in vegetative tissues (Zeevaart and Creelman, 1988). Thus mutants which are unable to synthesize ABA display a wilty phenotype, as in seedlings of the maize *vp* homozygotes, or tomato plants homozygous for the ABA-deficient *flacca* or *sitiens* mutations. The phenomenon has perhaps been best characterized in the responses of stomatal guard cells to water stress and ABA. In plants subjected to stress, high concentrations of ABA accumulate in the guard cells and, conversely, the application of ABA to leaves causes stomatal closure (Hartung and Davies, 1991). Experimental systems based on the responses of guard cells are proving extremely valuable for unravelling the nature of cellular and molecular responses of plants to ABA (Taylor *et al.*, 1995). Analysis of the response of vegetative plants to osmotic challenge has resulted in the development of a model to explain the molecular responses observed. Briefly stated, the model projects that the perception of stress results in the synthesis and accumulation of ABA, which in turn elicits a response at the level of gene expression, following the excitation of an intracellular signal transduction pathway.

This model does not appear to be applicable to developing embryos. First, the imposition of osmotic stress does not induce ABA biosynthesis. Measurements of endogenous ABA levels show that even when embryos are incubated under conditions which cause complete arrest of precocious germination, no significant changes in the endogenous concentrations of ABA can be detected (Morris *et al.*, 1990) However, it is clear that an ABA-related signal transduction pathway is implicated in the expression of both storage protein and *LEA* genes. Such conclusions can be drawn from the detailed molecular analyses of the expression of these genes which have been conducted in recent years. Thus, many of the ABA-induced *LEA* genes have been found to possess common *cis*-acting regulatory elements within their promoter regions, which confer ABA inducibility upon their associated coding sequences. Typically, these 'ABA responsive elements' (ABREs) contain a consensus sequence 'XACGTG'. Such sequences are not uncommon in plant genes (Williams *et al.*, 1992). Originally designated 'G-boxes' (to describe the palindromic sequence CCACGTGG), they are binding sites for transcription factors of the 'bZip' class — proteins which bind DNA through a stretch of basic amino acids, whilst forming a dimeric complex with an identical or related protein through a leucine-rich alpha-helix (the 'leucine zipper'). The specificity of the interaction may reside both in the sequence context in which the core sequence of the element resides, and in the components of the dimeric bZip complex (Schindler *et al.*, 1992; Williams *et al.*, 1992). The first such interaction to be described was that between the ABRE identified in the promoter of a wheat Em gene, and its cognate 'Em binding protein' (EMBP-1: Guiltinan *et al.*, 1990). Experiments utilizing fusions between Em promoter elements and the β-glucuronidase (GUS) reporter gene have demonstrated that the same promoter sequences that are required for ABA inducibility of gene expression are also required for induction by osmotic stress. The ABA/stress response has been observed in a number of experimental systems, including cereal protoplast systems (Marcotte *et al.*, 1988), in intact embryos following microprojectile bombardment (Öndé *et al.*, 1994), in transgenic dicotyledonous plants (Marcotte *et al.*, 1989), and even in transgenic bryophytes (Knight *et al.*, 1995). The extensive conservation of this response to both a plant growth regulator and a source of environmental stress argues strongly for an early evolutionary origin for the molecular response to osmotic stress during the colonization of terrestrial habitats by the first plants.

The interaction between the ABRE and the corresponding bZip protein is not in itself sufficient to elicit gene expression in response to ABA. This is clear from the results of studies of the expression of ABA-responsive genes in the viviparous mutants of maize, and that of their homologues in *Arabidopsis thaliana*. Thus, neither the endogenous globulin mRNA nor Em mRNA accumulates in developing maize embryos that are homozygous for any of the *vp* mutations (Butler and Cuming, 1993; Rivin and Grudt, 1991; Williams and Tsang, 1991). However, at least in the ABA-deficient mutants, the expression of these genes can be restored by supplementation of mutant embryos with exogenous ABA, a treatment which also prevents vivipary. In the *vp-1*, ABA-insensitive mutant, however, addition of ABA neither imposes dormancy nor induces globulin or Em gene expression. Furthermore, the application of osmotic stress does not induce expression of these genes, although it is able to prevent precocious germination through the imposition of a physical block to cell expansion (Butler and Cuming, 1993). The role of the *VP-1* gene in the regulation of gene expression was elucidated by McCarty, following the cloning of the *VP-1* gene by transposon tagging. In protoplast transfection experiments it was demonstrated that the *VP-1* gene product acted as a transcriptional activator mediating the expression of ABA-induced Em gene expression (McCarty *et al.*, 1989, 1991).

This identifies the *VP-1* gene as a component of the ABA-signal transduction chain, but a component which is situated towards the end of this chain. Moreover, its pattern of expression indicates that its activity is highly seed-specific. Further evidence for this view is provided by observations that, although *vp-1* mutant embryos are incapable of responding to applications of ABA in terms of following the embryonic maturation pathway leading to dormancy, they remain capable of responding to ABA in other ways. Thus, not all ABA-responsive *LEA* genes require a functional *VP-1* gene for their ABA- or stress-inducible expression. Whereas many of the cereal *LEA* genes are *VP-1* dependent (in maize, these include the Em, *glb-1* globulin and *rab 17* 'dehydrin') some genes remain induced by ABA treatment in *vp-1* embryos undergoing precocious germination. These include the *cat-1* gene (Williamson and Scandalios, 1992) and the *rab-28* gene (Pla *et al.*, 1991), in addition to a gene which encodes a protein designated 'MLG-3' — a group 3 *LEA* protein (Thomann *et al.*, 1992). Interestingly, although the *rab-28* gene is independent of *VP-1*, its stress- and ABA-mediated expression operates through a *cis*-acting ABRE sequence apparently identical to that of *VP-1* dependent genes (Pla *et al.*, 1993). In addition, comparisons of the patterns of proteins synthesized in *vp-1* mutants subjected to osmotic stress with those synthesized in mutant embryos treated with ABA reveal further differences (Butler and Cuming, 1993), indicating the existence of multiple stress- and ABA-response pathways.

Such a multiplicity of response networks appears to be a general phenomenon and has been observed not just in the case of genes which encode unrelated proteins, such as the genes that encode the Em polypeptide and the *rab 28* protein in maize, but also for genes which are closely related members of the same sequence family. For example, whereas maize appears to possess only a single DNA sequence encoding the Em polypeptide in its genome, other cereals possess different numbers of Em genes. Thus in hexaploid wheat there are 10–12 Em genes which appear to encode a highly homogeneous collection of polypeptide products that are indistinguishable by two-dimensional polyacrylamide gel electrophoresis (Futers *et al.*, 1990). By contrast, its diploid close relative barley (*Hordeum vulgare* L.) has been found to contain four Em genes, whose polypeptide products display variation in length due to the presence of a variably repeated 20-amino-acid domain. All of these genes are induced by ABA and by mannitol-induced stress, but interestingly their responses differ when an ionic stress is applied. One member of the family only appears to be responsive to salt stress (Espelund *et al.*,

1992; Hollung *et al.*, 1994). Such a complexity of embryonic stress and ABA responses is not restricted to cereal embryos. In *Arabidopsis thaliana*, different patterns of ABA- and stress-responsive gene expression have also been observed in ABA-insensitive mutant backgrounds (Parcy *et al.*, 1994). The amenability of this model species for genetic experimentation has resulted in the identification of a larger number of ABA-insensitive mutants than have been isolated in maize. To date, five separate ABA-insensitive mutant phenotypes have been described (designated *abi-1* to *abi-5*) (Finkelstein, 1994). Of these, the *ABI-3* gene has been cloned and shown to be a homologue of the maize *VP-1* gene (Giraudat *et al.*, 1992). The expression of a number of *LEA* proteins has been studied in *abi-3* mutants and shown to be similar to that observed in *vp-1* mutant maize embryos. Of the other genes, molecular characterization has so far been achieved only for the *ABI-1* gene. This has features which suggest that it acts as a protein phosphatase, but its mode of action has yet to be fully determined (Leung *et al.*, 1994; Meyer *et al.*, 1994). It is likely that this gene product is a component of an ABA signal transduction chain and acts at an earlier stage than *ABI-3/VP-1* in plant tissues. We can expect the further elucidation of the ABA response in *Arabidopsis* to be of crucial significance in furthering our understanding of the regulation of dormancy and maturation in cereals.

7.5 The expression of markers in the reprogrammed embryo

We have identified two points in the development of the embryo when the genetic programme appears to be fundamentally switched, the first of which initiates the onset of the maturation programme, to the exclusion of a germinative pathway, whilst the second is brought about by dehydration, to close the maturation pathway and initiate subsequent germination. To what extent can these developmental switches be regarded as irrevocable in the genetic programming of the cereal embryo? Insights into this question can be gained from studying the ability of embryos to express developmentally regulated genes at inappropriate times during development. We have described how the normal developmental sequence leading to the formation of a fully differentiated, mature embryo can be radically disrupted by the initiation of dedifferentiated callus tissue. This occurs when embryos isolated from immature grains at *c.* 12 dpa (the interface between stages 2 and 3) are incubated in the presence of 2,4-D. Under these conditions, a substantial mass of undifferentiated tissue forms within 1 week of callus induction. The genetic programme initiated by callus formation is not that typical of embryonic maturation. The expression of globulin and Em genes is suppressed, and instead there is a rapid accumulation of the germination-related mRNA encoding germin (Figure 7.4a). However, this occurs in the absence of any obvious germinative development (Figure 7.2), and it also takes place more rapidly within the 2,4-D-treated embryo than in a normal, precociously germinating embryo (cf. Figure 7.3d). These changes are not irreversible, and if callus tissue is treated with ABA, or subjected to osmotic stress during its formation, then the accumulation

(a) Germin mRNA

(b) Em mRNA

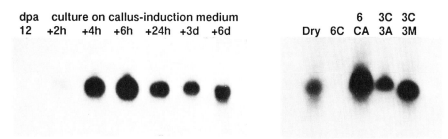

dpa   culture on callus-induction medium
12   +2h   +4h   +6h   +24h   +3d   +6d

6   3C   3C
Dry 6C   CA   3A   3M

Figure 7.4. *Expression of mRNA markers during dedifferentiation.*
(a) Germin mRNA. Total RNA (10 µg) was isolated from 12 dpa immature embryos incubated on MS medium containing 2,4-D at 2 mg l^{-1}. Germin mRNA is detectable by Northern hybridization within 2 h of incubation, and accumulates to very high levels by 6 h. The formation of callus becomes apparent only after c. 3 days of incubation. (b) Em mRNA. RNA was isolated from mature, dry embryos, 12 dpa embryos incubated for 6 days in the presence of 2,4-D (6C), 2,4-D + 10^{-5}M ABA (6CA) or for 3 days on 2,4-D followed by 3 days on medium containing 2,4-D and either 10^{-5}M ABA (3C3A) or 20% mannitol (3C3M).

of embryogenic markers, such as Em mRNA, occurs (Figure 7.4b). This is probably related to the continued ability of callus tissue derived in this way to undergo cellular redifferentiation and to regenerate via somatic embryogenesis (Maddock and Lancaster, 1983). Such cells are typically small and non-vacuolate. Callus tissue which is maintained for long periods on callus-inducing medium will continue to proliferate, but it loses the ability to express those markers characteristic of the differentiated state, and it is composed mainly of larger, highly vacuolate cells. This finding is correlated with a loss of ability to undergo subsequent somatic embryogenesis, suggesting that the expression of these embryogenesis- and germination-specific markers provides a measure of the embryogenic capacity of tissue culture. A similar phenomenon has been observed in gymnosperm embryo-derived cultures, in which the germin homologues have also been shown to be expressed upon the initiation of callus formation, and to be expressed only in embryogenic cultures (Domon *et al.*, 1995). These observations may be related to the enzymatic activity characteristic of germin, namely the activity of oxalate oxidase, an enzyme which generates hydrogen peroxide as a reaction product. This highly reactive compound is known to be required for cross-linking reactions in the cell wall (Fry, 1986), and the correlation of peroxide-generating enzyme activity with the embryogenic state may indicate a requirement for maintaining the population of small, densely cytoplasmic cells from which somatic embryogenesis may be initiated. A similar proposal has been made to explain the association of peroxidases with the embryogenic state in carrot suspension cultures (Chapter 11, Toonen and de Vries).

Germination itself is not sufficient to close the embryogenic programme irreversibly following the imbibition of water by dry embryos. Expression of both

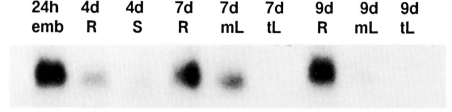

| 24h | 4d | 4d | 7d | 7d | 7d | 9d | 9d | 9d |
| emb | R | S | R | mL | tL | R | mL | tL |

Figure 7.5. *Stress-induced Em mRNA expression in seedlings. Whole grains were germinated and seedlings grown in hydroponic culture for periods of 24 h to 9 days. The plants were then transferred to hydroponic medium containing 20% mannitol for a further 24 h, prior to the isolation of total RNA. RNA was isolated from whole embryos (24 h), from separated roots (R) and shoots (S) (4d), and from roots (R), coleoptile + mid-leaf (mL) and leaf tips (tL) of older seedlings (7d and 9d).*

storage proteins and *LEA* gene products has been shown to be recapitulated in germinating embryos and young seedlings following subjection to stress or treatment with ABA (Morris *et al.*, 1990; Mundy and Chua, 1988). In wheat seedlings subjected to osmotic stress by the inclusion of mannitol in the growth medium, the mRNA encoding the Em protein accumulates even in well-established plants (9-day-old seedlings: Figure 7.5). Levels of expression in these seedlings appear to be higher in the roots than in the aerial parts of the plant, perhaps because these organs are in direct contact with the medium, and so are subjected to a higher degree of stress, or because stress brings about a reallocation of water within the plant. Such an explanation has been proposed to explain similar observations of the expression of a salt-induced gene (*salT*) in rice seedlings (Claes *et al.*, 1990). Finally, the expression of these genes can also be initiated in aleurone cells. This tissue has long been studied as a target tissue for the action of gibberellins and ABA in seed development and germination, and aleurone cells can be used as vehicles for the transient expression of genes under the control of Em and other *LEA* gene promoters, in response to ABA (Jacobsen and Close, 1991; Öndé *et al.*, 1994; Robertson *et al.*, 1995). The fact that apparently 'embryo-specific' genes can be expressed in this endosperm-derived tissue may reflect the suggestion made in Chapter 10, that the endosperm represents a 'reprogrammed embryo'.

References

Bennett, M.D., Rao, M.K., Smith, J.B. and Bayliss, M.W. (1973) Cell development in the anther, the ovule, and the young seed of *Triticum aestivum* L. var Chinese Spring. *Phil. Trans. Roy. Soc. Lond.* **B266**, 6–81.

Butler, W.M. and Cuming, A.C. (1993) Differential molecular responses to abscisic acid and osmotic stress in viviparous maize mutants. *Planta* **189**, 47–54.

Claes, B., DeKeyser, R., Villaroel, R., Van den Bulcke, M., Bauw, G., Van Montagu, M. and Caplan, A. (1990) Characterization of a rice gene showing organ specific expression in response to salt stress and drought. *Plant Cell* **2**, 19–27.

Close, T.J., Kortt, A.A. and Chandler, P.M. (1989) A cDNA-based comparison of dehydration induced proteins (dehydrins) in barley and corn. *Plant Mol. Biol.* **13**, 95–108.

Crowe, J.H. and Crowe, L.M. (1992) Membrane integrity in anhydrobiotic organisms: toward a mechanism for stabilizing dry cells. In: *Water and Life* (eds G. Somero, C.B. Osmond and C.L. Bolis). Springer-Verlag, Berlin, pp. 87–103.

Cuming, A.C. (1984) Developmental regulation of gene expression in wheat embryos. Molecular cloning of a DNA sequence encoding the early-methionine-labelled (Em) polypeptide. *Eur. J. Biochem.* **145**, 351–357.

Domon, J.M., Dumas, B., Laine, E., Meyer, Y., David, A. and David, H. (1995) 3 glycosylated polypeptides secreted by several embryogenic-cell cultures of pine show highly specific seriological affinity antibodies directed against the wheat germin apoprotein monomer. *Plant Physiol.* **108**, 141–148.

Dumas, B., Sailland, A., Cheviet, J.P., Freyssinet, G. and Pallett, K. (1993) Identification of barley oxalate oxidase as a germin-like protein. *Comptes Rendus Acad. Sci. Paris* **316**, 793–798.

Dure, L.S. III (1993) A repeating 11-mer amino acid motif and plant desiccation. *Plant J.* **3**, 363–369.

Edsall, J.T. and McKenzie, H.A. (1983) Water and proteins. II. The location and dynamics of water in protein systems and its relation to their stability and properties. *Adv. Biophys.* **16**, 53–183.

Espelund, M., Sæbøe-Larssen, S., Hughes, D.W., Galau, G.A., Larsen, F. and Jakobsen, K.S. (1992) Late embryogenesis-abundant genes encoding proteins with different numbers of hydrophilic repeats are regulated differentially by abscisic acid and osmotic stress. *Plant J.* **2**, 241–252.

Finkelstein, R.R. (1994) Mutations at two new *Arabidopsis* ABA response loci are similar to the *abi3* mutations. *Plant J.* **5**, 765–771.

Fry, S.C. (1986) Cross-linking of matrix polymers in the growing cell walls of angiosperms. *Ann. Rev. Plant Physiol.* **37**, 165–186.

Futers, T.S., Vaughan, T.J., Sharp, P.J. and Cuming, A.C. (1990) Molecular cloning and chromosomal location of genes encoding the 'Early methionine labelled' (Em) polypeptide of *Triticum aestivum* L. var. Chinese Spring. *Theor. Appl. Genet.* **80**, 43–48.

Galau, G.A., Hughes, D.W. and Dure, L.S. III (1986) ABA induction of cloned cotton late embryogenesis-abundant (LEA) mRNAs. *Plant Mol. Biol.* **7**, 155–170.

Giraudat, J., Hauge, B.M., Valon, C., Dmalle, J., Parcy, F. and Goodman, H.M. (1992) Isolation of the *Arabidopsis ABI3* gene by positional cloning. *Plant Cell* **4**, 1251–1261.

Grzelczak, Z.F. and Lane, B.G. (1984) Signal resistance of a soluble protein to enzymic proteolysis. An unorthodox approach to the isolation and purification of germin, a rare growth-related protein. *Can. J. Biochem. Cell Biol.* **62**, 1351–1353.

Grzelczak, Z.F., Sattolo, M.H., Hanley-Bowdoin, L.K., Kennedy, T.D. and Lane, B.G. (1982) Synthesis and turnover of proteins and messenger RNA in germinating wheat embryos. *Can. J. Biochem.* **60**, 389–397.

Guiltinan, M.J., Marcotte, W.R. and Quatrano, R.S. (1990) A plant leucine zipper protein that recognises an abscisic acid response element. *Science* **250**, 267–270.

Hartung, W. and Davies, W.J. (1991) Drought induced changes in physiology and ABA. In: *Abscisic Acid: Physiology and Biochemistry* (eds W.J. Davies and H.G. Jones). BIOS Scientific Publishers, Oxford, pp. 63–80.

Hilhorst, H.W.M. (1995) A critical update on seed dormancy. I. Primary dormancy. *Seed Science Res.* **5**, 61–74.

Hollung, K., Espelund, M. and Jakobsen, K.S. (1994) Amino-acid hydrophilic motif. *Plant Mol. Biol.* **25**, 559–564.

Iturriaga, G., Schneider, K., Salamini, F. and Bartels, D. (1992) Expression of desiccation-related proteins from the resurrection plant *Craterostigma plantagineum* in transgenic tobacco. *Plant Mol. Biol.* **20**, 555–558.

Jacobsen, J.V. and Close, T.J. (1991) Control of transient expression of chimaeric genes by gibberellic acid and abscisic acid in protoplasts prepared from mature barley aleurone layers. *Plant Mol. Biol.* **16**, 713–724.

Knight, C.D., Sehgal, A., Atwal, K., Wallace, J.C., Cove, D.J., Coates, D., Quatrano, R.S., Bahadur, S., Stockley, P.G. and Cuming, A.C. (1995) Molecular responses to abscisic acid and stress are conserved between moss and cereals. *Plant Cell.* **7**, 499–506.

Lane, B.G. (1991) Cellular desiccation and hydration: developmentally regulated proteins and the maturation and germination of cereal embryos. *FASEB J.* **5**, 2893–2901.

Lane, B.G., Bernier, F., Dratewka-Kos, E., Shafai, R., Pyne, C., Munro, J.R., Vaughan, T., Walters, D. and Altomare, F. (1991) Homologies between members of the germin gene famoil in hexaploid wheat and similarities between these wheat germins and certain *Physarum* spherulins. *J. Biol. Chem.* **266**, 10461–10469.

Lane, B.G., Dunwell, J.M., Ray, J.A., Schmitt, M.R. and Cuming, A.C. (1993) Germin, a protein marker of early plant development, is an oxalate oxidase. *J. Biol. Chem.* **268**, 12239–12242.

Leopold, A.C., Bruni, F. and Williams, R.J. (1992) Water in dry organisms. In: *Water and Life* (eds G. Somero, C.B. Osmond and R.J. Williams). Springer-Verlag, Berlin, pp. 161–169.

Leung, J., Bouvier-Durand, M., Morris, P.C., Guerrier, D., Chefdor, F. and Giraudat, J. (1994) *Arabidopsis* ABA response gene ABI1 — features of a calcium-modulated protein phosphatase. *Science* **264**, 1448–1452.

McCarty, D.R., Carson, C.B., Stinard, P.S and Robertson, D.S. (1989) Molecular analysis of viviparous-1: an abscisic acid insensitive mutant of maize. *Plant Cell* **1**, 523–532.

McCarty, D.R., Hattori, T., Carson, C.B., Vasil, V., Lazar, M. and Vasil, I.K. (1991) The *viviparous-1* developmental gene of maize encodes a novel transcriptional activator. *Cell* **66**, 895–905.

McCubbin, W.D., Kay, C.M. and Lane, B.G. (1985) Hydrodynamic and optical properties of the wheat germ *Em* protein. *Can. J. Biochem.* **63**, 803–811.

McKersie, B.D. and Senaratna, T. (1983) Membrane structure in germinating seeds. *Recent Adv. Phytochem.* **17**, 29–52.

Maddock, S.E. and Lancaster, V.A. (1983) Plant regeneration from cultured immature embryos and inflorescences of 25 cultivars of wheat (*Triticum aestivum*). *J. Exp. Bot.* **34**, 915–926.

Marcotte, W.R. Jr, Bayley, C.C. and Quatrano, R.S. (1988) Regulation of a wheat promoter by abscisic acid in rice protoplasts. *Nature* **335**, 454–457.

Marcotte, W.R. Jr, Russell, S.H. and Quatrano, R.S. (1989) Abscisic acid-responsive sequences from the Em gene of wheat. *Plant Cell* **1**, 969–976.

Meyer, K., Leube, M.P. and Grill, E. (1994) A protein phosphatase 2C involved in ABA signal-transduction in *Arabidopsis thaliana*. *Science* **264**, 1452–1455.

Morris, P.C., Kumar, A., Bowles, D.J. and Cuming, A.C. (1990) Osmotic stress and abscisic acid regulate expression of the wheat *Em* genes. *Eur. J. Biochem.* **190**, 625–630.

Mundy, J. and Chua, N.H. (1988) Abscisic acid and water stress induce the expression of a novel rice gene. *EMBO J.* **7**, 2279–2286.

Neill, S.J., Horgan, R. and Parry, A.D. (1986) The carotenoid and abscisic acid content of viviparous kernels and seedlings of *Zea mays* L. *Planta* **169**, 87–96.

Neill, S.J., Horgan, R. and Rees, A.F. (1987) Seed development and vivipary in *Zea mays* L. *Planta* **171**, 358–364

Öndé, S., Futers, T.S. and Cuming, A.C. (1994) Rapid analysis of an osmotic stress responsive promoter by transient expression in intact wheat embryos. *J. Exp. Bot.* **45**, 561–566.

Parcy, F., Valon, C., Raynal, M., Gaubier-Comella, P., Delseny, M. and Giraudat, J. (1994) Regulation of gene expression programs during *Arabidopsis* seed development: roles of the ABI3 locus and of endogenous abscisic acid. *Plant Cell* **6**, 1567–1582.

Pla, M., Gomez, J., Goday, A. and Pages, M. (1991) Regulation of the abscisic acid responsive gene *rab-28* in maize *viviparous* mutants. *Mol. Gen. Genet.* **230**, 394–400.

Pla, M., Vilardell, J., Guiltinan, M.J., Marcotte, W.R., Niogret, M.F., Quatrano, R.S. and Pages, M. (1993) The *cis*-regulatory element CCACGTGG is involved in the ABA and water-stress responses of the maize gene *rab 28*. *Plant. Mol. Biol.* **21**, 259–266.

Quatrano, R.S., Litts, J., Colwell, G., Chakerian, R. and Hopkins, R. (1986) Regulation of gene expression in wheat embryos by ABA: characterisation of cDNA clones for *Em* and putative globulin proteins and localisation of the lectin wheat germ agglutinin. In: *Molecular Biology of Seed Storage Proteins and Lectins* (eds L.M. Shannon and M.J. Chrispeels). American Society of Plant Physiologists, Rockville MD, pp. 127–136.

Rivin, C.J. and Grudt, T. (1991) Abscisic acid and the developmental regulation of embryo storage proteins in maize. *Plant Physiol.* **95**, 358–365.

Rhodes, D. and Hanson, A.D. (1993) Quarternary ammonium and tertiary sulfonium compounds in higher plants. *Ann. Rev. Plant Physiol. Plant Mol. Biol.* **44**, 357–384.

Robertson, D.S. (1955) The genetics of vivipary in maize. *Genetics* **40**, 745–760.

Robertson, M., Cuming, A.C. and Chandler, P.M. (1995) Sequence analysis and hormonal regulation of a dehydrin promoter from barley, *Hordeum vulgare* L. 'Himalaya'. *Physiol. Plant.* **94**, in press.

Rogers, S.O. and Quatrano, R.S. (1983) Morphological staging of wheat caryopsis development. *Am. J. Bot.* **70**, 308–311.

Schindler, U., Menkens, A.E., Beckmann, H., Ecker, J.R. and Cashmore, A.R. (1992) Heterodimerisation between light-regulated and ubiquitously expressed GBF bZip proteins. *EMBO J.* **11**, 1261–1273.

Taylor, J.E., Renwick, K.F., Webb, A.A.R., McAinsh, M.R., Furini, A., Bartels, D., Quatrano, R.S., Marcotte, W.R. Jr and Hetherington, A. (1995) ABA regulated promoter activity in stomatal guard cells. *Plant J.* **7**, 129–134.

Thomann, E.B., Sollinger, J., White, C. and Rivin, C.J. (1992) Accumulation of group 3 late embryogenesis abundant proteins in *Zea mays* embryos — roles of abscisic acid and the *viviparous-1* gene product. *Plant Physiol.* **99**, 607–614.

Türet, M. (1993) Molecular analysis of changes in gene expression during callus induction from wheat embryos. Ph.D. Thesis, University of Leeds.

Vernon D.M. and Bohnert H.J. (1992) A novel methyl transferase induced by osmotic stress in the facultative halophyte *Mesembryanthemum crystallinum*. *EMBO J.* **11**, 2077–2085.

Williams, B. and Tsang, A. (1991) A maize gene expressed during embryogenesis is abscisic acid inducible and highly conserved. *Plant Mol. Biol.* **16**, 919–923

Williams, M.E., Foster, R. and Chua, N.-H. (1992) Sequences flanking the G-box core CACGTG affect the specificity of protein binding. *Plant Cell* **4**, 485–496.

Williamson, J.D. and Scandalios, J. (1992) Differential response of maize catalases to abscisic acid: Vp1 transcriptional activator is not required for abscisic acid-regulated Cat1 expression. *Proc. Natl. Acad. Sci. USA* **89**, 8842–8846.

Yancey, P., Lark, M., Hand., S., Bowlus, R. and Somero, G. (1982) Living with water stress: evolution of osmolyte systems. *Science* **217**, 1214–1222.

Zeevaart, J.A.D. and Creelman, R.A. (1988) Metabolism and physiology of abscisic acid. *Ann. Rev. Plant Physiol Plant Mol. Biol.* **39**, 439–473.

8

Genetic analysis of seed maturation and germination pathways in maize

D.R. McCarty

8.1 Introduction

The viviparous mutants of maize have provided an extraordinarily rich source of insight into the mechanisms which control seed maturation and germination in plants, recently reviewed by McCarty (1995). Most of the well-characterized maize mutants affect aspects of abscisic acid (ABA) hormone synthesis and perception in the developing seed (Neill et al., 1986, 1987). However, our understanding of the underlying mechanisms which integrate hormonal signals with intrinsic developmental pathways is far from complete.

The gene for the ABA-insensitive *vp1* mutant has been cloned and studied extensively at the molecular level (Hattori *et al.*, 1992; McCarty *et al.*, 1989, 1991; Vasil *et al.*, 1995). Functional analysis of the *Vp1* protein indicates that it is a novel transcriptional activator which interacts synergistically with ABA-egulated transcription factors (Hattori *et al.*, 1992; McCarty *et al.*, 1991; Vasil *et al.*, 1995). However, key aspects of the *vp1* phenotype and the related *abi3* mutant of *Arabidopsis* suggest that *vp1/Abi3* action is not limited to ABA signal transduction (for a discussion of these issues see McCarty, 1992, 1995; Parcy *et al.*, 1994). Moreover, the *fusca3* (Keith *et al.*, 1994) and *lec1* (Meinke *et al.*, 1995) viviparous mutants of *Arabidopsis* provide additional evidence for an ABA-independent cascade regulating seed maturation. The *slender* mutants of barley (Chandler, 1988; Lanahan and Ho, 1988) and the *spindly* mutants of *Arabidopsis* (Olszewski and Jacobsen, 1993) indicate that mutation in the gibberellic acid (GA) signalling pathway may also affect the control of germination.

The mechanisms underlying developmental control of ABA synthesis are also poorly understood. In the majority of the existing ABA-deficient mutants of maize, the early steps in the pathway for biosynthesis of the carotenoid precursors of ABA are blocked (Neill *et al.*, 1986). This class of pleiotropic albino mutants can provide little insight into the developmental control of ABA synthesis in plants. However, there is ample evidence for a more maturation-specific class of viviparous mutants (McCarty, 1995).

In order to address these issues, we have attempted a comprehensive genetic analysis of the viviparous mutants, with the emphasis being placed on mutants that are likely to affect specifically the control of maturation and/or germination. We have concentrated on transposon-induced mutations to facilitate the eventual cloning and molecular analysis of these mutants.

8.2 Transposon mutagenesis

The *Robertson's Mutator* transposable element system has proved widely successful in mutagenesis and gene-tagging studies in maize (Walbot, 1992). *Robertson's Mutator* has the advantage of producing a high mutation frequency in active lines (typically of the order of 10^{-4}) with a wide spectrum of phenotypic classes. *Mutator* has been used effectively both for targeted tagging strategies directed at finding new tagged alleles of known mutants (e.g. mutants previously identified by chemical mutagenesis), and for non-targeted strategies aimed at simultaneous discovery and tagging of new genes in the pathway of interest. In a targeted screen, tagged alleles are identified as rare mutant individuals in a large F_1 population generated by crossing a *Mutator* line to a non-*Mutator* tester line which carries the target mutant. Because recovery of the newly tagged allele in this situation depends on the ability to obtain seed from homozygous mutant individuals identified in the initial screen, targeted strategies are not feasible for screens that involve potentially lethal phenotypes such as the viviparous seed mutations. An alternative non-targeted approach involves screening a large number of lines derived by self-pollinating active *Mutator* lines for ears which segregate viviparous seed. On ears which segregate 3:1 for a recessive viviparous mutation, two-thirds of the normal seed will be heterozygous for the new mutation. This allows propagation of the new mutant as a heterozygote. The new mutant can be made homozygous and studied as required by self-pollinating heterozygotes.

8.2.1 *Genetic stocks*

The mutants described in the present study were derived from a collection of 25 putatively viviparous lines assembled by Dr Philip Stinard in a survey of self-pollinated *Robertson's Mutator* ears grown by Dr Donald S. Robertson at Iowa State University between 1985 and 1989. The *vp1*-R, *vp1*-w2 and wild type (*A1*, *A2*, *Bz1*, *Bz1*, *C1*, *C2* and *R1*) maize lines were in a W22 inbred background. The *vp10* stock was obtained from the Maize Genetic Cooperative Stock Center (Urbana, IL, USA).

8.3 Mutator-induced viviparous mutants

In the majority of the classical viviparous mutants of maize, the synthesis of the carotenoid precursors of abscisic acid is blocked. Because carotenoid-deficient mutants only indirectly affect developmental control of the maturation pathway, we selected only those viviparous mutants which formed green (chlorophyll-containing) shoots in the mutant seed for further analysis. In addition, available pedigree records were used to identify lines that were likely to be derived from the same mutational event. On the basis of these criteria, we identified 17 independent lines that were subsequently confirmed as carrying heritable, green embryo-type viviparous mutants (Table 8.1). In order to detect the effects of the new mutants on anthocyanin synthesis and to determine F_2 segregation ratios, the selected lines were outcrossed to a W22 inbred line with full aleurone anthocyanin pigmentation (genotypes *A1*, *A2*, *Bz1*, *Bz2*, *C1*, *C2* and *R1*). The resulting progeny were then self-pollinated in order to recover the segregating

Table 8.1. Robertson's Mutator *transposon-induced viviparous mutants in maize*

Complementation group	Accession	F_2 segregation	Anthocyanin phenotype	Penetrance
Group 0: *vp1*				
	vp*-2327	3:1	Mottled, variable	Mu suppressible
	vp*-5315	3:1	Deficient, spotted	Strong
	vp*-2299	3:1	Deficient, spotted	Strong
Group 1: *vp13*				
	vp*-1126	3:1	Normal	Strong
	vp*-3040	3:1	Normal	Strong
	vp*-3339	3:1	Normal	Strong
	vp*-5279	3:1	Normal	Strong
Group 2: (*vp10?*)				
	vp*-2274	3:1	Normal	variable
	vp*-3286	3:1	Normal	Weak, variable
	vp*-3250	3:1	Normal	Weak, variable
Group 3:				
	vp*-2406	15:1	Normal	Strong
	vp*-2339	15:1	Normal	Strong
Unassigned:				
	vp*-1379	n.d.[a]	Normal	Moderate
	vp*-2458	3:1	Normal	Strong
	vp*-3017	n.d.	Normal	Moderate
	vp*-3239	3:1	Normal	Strong
	vp*-1409	?	Normal	Very weak

[a]n.d., not determined.

F$_2$ seed. In addition, plants from each segregating line were pollinated by homozygous mutant *vp1*-R or *vp1*-w2 plants in order to test for complementation with the *vp1* mutant.

We identified three viviparous mutants (*vp**-2327, *vp**-5315 and *vp**-2299) that co-segregated with an anthocyanin-deficient aleurone phenotype and also failed to complement the *vp1* mutant. These new alleles of *vp1* are designated *vp1*-mum3, *vp1*-mum4 and *vp1*-mum5, respectively. All of the 14 remaining mutants complemented the *vp1* mutant and exhibited normal anthocyanin pigmentation in the aleurone. To date, *vp1* remains the only known mutant of maize that both affects anthocyanin pigmentation and causes vivipary.

8.3.1 *Identification and mapping of a new viviparous locus*

The above crossing studies revealed a weak linkage between the *vp**-1126 mutation and the *R1* anthocyanin gene (approximately 20% recombination), indicating that this mutant was also likely to be located on chromosome 10. This chromosomal location was confirmed by fertilizing *vp**-1126 heterozygous plants with pollen from plants that were hyperploid for the T-B10S or T-B10L19 translocation chromosomes.

In maize, B-A translocation chromosomes typically undergo non-disjunction at the second mitotic division in pollen grain development to generate genetically non-identical sperm cells in the pollen grain. As a result, pollination produces two classes of genetically non-concordant seed: a class in which the embryo carries two copies of the translocated A chromosome segment and the endosperm is deficient for that segment, and the reciprocal class in which the endosperm is hyperploid for the A chromosome arm and the embryo is deficient (hypoploid). If the female parent carries an embryo mutation in the affected region of the chromosome, the hypoploid embryo class will exhibit the mutant phenotype in the F$_1$ generation. In the case of the *vp**-1126 mutation, crosses involving the T-B10L19 translocation chromosome uncovered the viviparous phenotype in the hypoploid embryo class, whereas the T-B10S cross yielded only normal seed. This indicated that *vp**-1126 is located on the long arm of chromosome 10, consistent with the evidence for linkage to the *R1* locus.

No previously described viviparous mutations map to chromosome 10, indicating that *vp**-1126 has identified a new locus. On this basis, we have designated the new locus *vp13*.

8.3.2 *Complementation analysis*

In order to resolve the mutants further into allelic groups, the mutant lines were systematically intercrossed and tested for complementation. The complementation groups assigned to date are summarized in Table 8.1. Four alleles of the *vp13* mutant have been confirmed (now designated *vp13*-mu1126, *vp13*-mu3040, *vp13*-mu5279 and *vp13*-mu3339). The *vp**-2274, *vp**-3250 and *vp**-3286 mutants represent a second complementation group. In addition, preliminary test-cross results (D. R. McCarty, unpublished data) indicate that the vp^*-2274 mutant fails to complement the $vp10$ mutant stock (Smith and Neuffer, 1992) obtained from the Maize Cooperative Stock Center, allowing us to assign these alleles tentatively to the $vp10$ locus.

A third confirmed complementation group consists of the vp^*-2406 and vp^*-2339 alleles. Both of these mutants produced 15:1 (normal:mutant) segregation ratios in F_2 materials, suggesting the involvement of a duplicate factor. Consistent with a two-gene model, both 3:1 segregation ratios and 15:1 ratios were observed on F_3 ears. Although all of the currently known viviparous mutants segregate as single genes, in his early studies of viviparous mutants Manglesdorf (1930) also reported evidence of duplicate factor inheritance. Manglesdorf's mutant collection is unfortunately no longer extant. On the basis of these findings, we therefore suggest that vp^*-2406 and vp^*-2339 identify a pair of duplicate genes which control seed maturation.

A number of duplicate factors have been described in other pathways in maize (Cone *et al.*, 1993; Helentjaris *et al.*, 1988). This observation and the fact that large segments of the genome are duplicated in the restriction fragment length polymorphism (RFLP) map has been regarded as evidence for a tetraploid ancestry of maize (Helentjaris *et al.*, 1988). In the other cases that have been studied, the duplicated genes in maize have diverged functionally to widely varying degrees (e.g. Cone *et al.*, 1993). While the duplicate factor inheritance is an indication that the vp^*-2406 gene pair display functional overlap in the developing seed, it has yet to be determined whether these factors are fully redundant in all tissues. The evidence that we recovered two independent alleles in this group is remarkable, given the limited scope of the search for mutants and the expected rarity of mutations occurring in both genes in these lines. The most probable explanation is that the *Robertson's Mutator* materials used in this study carried a pre-existing mutation in one of the two genes.

The five remaining mutants (vp^*-1379, vp^*-1409, vp^*-2458, vp^*-3239 and vp^*-3017) represent a catch-all class and to date have not been confirmed as belonging to any of the above groups. We have not ruled out the possibility that these mutants identify additional complementation groups.

Overall, the results of the complementation analysis provide evidence for at least three genes in addition to $vp1$ which function specifically in the control of the seed maturation and germination pathways. The 17 mutants analysed included three $vp1$ alleles, at least four $vp13$ alleles, three putative $vp10$ mutants and two mutants in the duplicate factor class. These mutation frequencies, which are fairly well balanced across these loci, indicate that the pathway has not yet been fully saturated with mutants. On the other hand, the statistics suggest that we should not expect to find a large number of additional genes using this screen.

Our ability to clone and confirm the identity of the transposon-tagged mutants is greatly aided by the availability of multiple independent alleles of each locus. First, multiple alleles increase the probability of success in identifying a co-segregating *Mu* element in at least one mutant line. Secondly, cross-referencing between independent alleles is essential for confirmation of the identity of any candidate clones. Segregation analysis of these mutants is currently in progress.

8.3.3 *Penetrance and genetic background effects*

Much of the difficulty and ambiguity still surrounding our efforts to classify these mutants into complementation groups stems from the enigmatic phenotypic expression of some of the viviparous lines. Overall, the *vp13* alleles and *vp*-2406/2339* group have strongly penetrant phenotypes and produce clear segregation ratios. On the other hand, the putative alleles of *vp10* (*vp*-2274* and alleles) are much less penetrant and more variable in expression. Within this group, the mutant phenotype can range from a fully viviparous, desiccation-intolerant seed to a subtle elongation of the shoot axis in the mutant embryo. In the latter situation, mutant seed are typically desiccation tolerant and capable of germination. Although the mutants of this group form an approximately allelic series with *vp*-2274* having the strongest penetrance and *vp*-3286* the weakest, all three alleles display substantial phenotypic variation. For these reasons, the assignment of allelic relationships within this group in particular can be tedious.

We are investigating several factors which may possibly contribute to the phenotypic variation of these mutants. One intriguing possibility is that other related genes in the background with overlapping functions partially suppress the mutant gene. Complex differentially regulated gene families are an emerging theme in plant developmental pathways. For example, the genes which encode aminocyclopropane synthase, the key regulatory step in ethylene hormone synthesis, comprise a differentially regulated gene family in plants (Liang *et al.*, 1992). By analogy, we might expect the genes responsible for developmental control of the ABA biosynthetic pathway to be encoded by a gene family of similar or greater complexity. As discussed above, we have obtained direct genetic evidence for duplicate genes in the *vp*-2406* complementation group. Consistent with this scenario, preliminary molecular analysis of the *vp*-2274* gene-specific clones indicates that this may also be a multicopy gene (B.-C. Tan and D. R. McCarty, unpublished data).

A second factor that might contribute to variable phenotypic expression among alleles is the phenomenon of *Mutator* suppression (Barkan and Martienssen, 1988). Some *Mutator*-induced mutations become phenotypically suppressed in lines where *Mutator* transposition has become inactive, even though the transposon remains inserted in the gene. In the case of the *hcf*106 mutant which has been studied in detail, suppression is attributed to the proximity of the *Mu1* element to the normal transcription initiation site of the gene. Inactivation of *Mu1* transposition allows transcriptional initiation to occur from within the end of *Mu1*. Molecular studies have indicated that this mechanism is responsible for the suppressible phenotype of the *vp1*-mum3 (*vp*-2327*) mutant as well (R. Martienssen, personal communication; see Table 8.1).

Ultimately, the degree to which phenotypic variation can be attributed to background effects or 'leakiness' of the mutations will be determined by molecular analysis of these alleles.

8.3.4 *Characteristic phenotypes of the* vp13 *and* vp*-2274 *mutants*

With the exception of the allelic variation discussed above, the mutants within each complementation group exhibited similar phenotypes. In the *vp13* mutants the mutant embryo shoot axis is typically greatly elongated by the time the ear reaches maturity, but the embryo does not usually expand with sufficient force to rupture the surrounding pericarp. As a result, the elongating shoot forms a coil beneath the pericarp at the top of the kernel, giving the mutant seed a distinctive 'green turban' appearance. The extent of radicle elongation and emergence is highly variable. Mutant seed can be rescued by manually opening or removing the restraining pericarp and placing the seed in soil prior to desiccation of the seed. The resulting seedlings are narrow and spindly, and produce 5 to 6 leaves before becoming necrotic. With the emergence of the fourth or fifth leaf the elongation rate slows down and becomes irregular, causing some twisting and distortion of the emerging leaf. Necrosis is first evident at the tips of the uppermost leaves, and then progresses down the plant. There is little evidence of wilting or water stress that might indicate an ABA deficiency. In any case, topical applications of ABA did not slow down the progression of seedling necrosis (C. B. Carson and D. R. McCarty, unpublished data). In many respects, the seedling-lethal phenotype of the *vp13* mutant is strikingly similar to that of the non-extant *vp6* mutant described by Robertson (1955), suggesting that we may have rediscovered the earlier mutant. Overall, this phenotype suggests a disruption of shoot elongation control mechanisms in the *vp13* mutant. This aspect of the *vp13* phenotype is reminiscent of the barley *slender* mutant (Lanahan and Ho, 1988) and the *spindly* mutant of *Arabidopsis* (Olszewski and Jacobsen, 1993), which also affect control of seed germination.

In contrast to *vp13*, the *vp*-2274 type mutants (group 2) produce fully viable seedlings and can be maintained in the homozygous mutant condition. The viviparous seeds develop in a manner similar to normal germination and to other viable viviparous mutants, including *vp1* and *vp10* (Smith and Neuffer, 1992). However, as mentioned above, the mutant seed are not always viviparous or desiccation intolerant. The most consistently penetrant feature of the phenotype that distinguishes mutant from normal seeds is the elongation of the embryo shoot axis beyond the tip of the scutellum. Homozygous mutant plants are typically slightly shorter than their normal siblings at the time of flowering. Under the stressful conditions typical of the Florida growing season, the mutant plants do not differ dramatically from their normal siblings in susceptibility to water stress. However, this characteristic has been highly variable in the materials examined to date. Therefore, subtle or background-dependent effects of the mutant on plant water relations cannot be ruled out.

These characteristics, in addition to evidence that mutant *vp*-2274* embryos respond to exogenous ABA in culture (B.-C. Tan and D. R. McCarty, unpublished data), are consistent with the hypothesis that *vp*-2274* functions in a seed-specific ABA biosynthetic pathway.

8.4 Conclusions

Transposon mutagenesis using the *Robertson's Mutator* system has led to the identification of at least three new mutants in the pathway which controls seed maturation in maize. The distribution of alleles suggests that the screen based on selection of green viviparous embryo mutants is approaching but has not yet reached the point of statistical saturation. The identification of at least one duplicate factor and evidence for background effects on expression of several mutants in the collection suggests that redundant or overlapping expression of genes in the pathway may be constraining our ability to find new mutants in this pathway. In a broader sense, this suggestion is consistent with evidence that differentially regulated gene families play a major role in the developmental control of pathways in plants. The phenotypes of the newly identified *vp13* mutant and the putative *vp10* alleles suggest that different mechanisms underlie vivipary in each case. The proposed mechanisms include altered control of cell elongation and expansion and blocked ABA synthesis in the seed, respectively.

Acknowledgements

This work was supported by grants from the USDA Competitive Grants Program and the National Science Foundation, and by the Florida Agricultural Experiment Station. This chapter represents Journal Series Number R-04768.

References

Barkan, A. and Martienssen, R. (1988) Inactivation of maize transposon *Mu* suppresses a mutant phenotype by activating an outward-reading promoter near the end of *Mu1*. *Proc. Natl Acad. Sci. USA* **88**, 3502–3506.

Chandler, P.M. (1988) Hormonal regulation of gene expression in the *slender* mutant of barley (*Hordeum vulgare* L.). *Planta* **175**, 115–120.

Cone, K.C., Cocciolone, S.M., Moehlenkamp, C.A., Weber, T., Drummond, B.J., Tagliani, L.A., Bowen, B.A. and Perrot, G.H. (1993) Role of the regulatory gene *pl* in the photocontrol of maize anthocyanin pigmentation. *Plant Cell* **5**, 1807–1816.

Hattori, T., Vasil, V., Rosenkrans, L., Hannah, L.C., McCarty, D.R. and Vasil, I.K. (1992) The *viviparous-1* gene and abscisic acid activate the *C1* regulatory gene for anthocyanin biosynthesis during seed maturation in maize. *Genes Dev.* **6**, 609–618.

Helentjaris, T., Weber, D. and Wright, S. (1988) Identification of the genomic locations of duplicate neucleotide sequences in maize by analysis restriction fragment length polymorphisms. *Genetics* **118**, 353–363.

Keith, K., Kraml, M., Dengler, N.G. and McCourt, P. (1994) *Fusca3:* a heterochronic mutation affecting late embryo development in *Arabidopsis. Plant Cell* **6**, 589–600.

Lanahan, M.B. and Ho, T.D. (1988) Slender barley: a constitutive gibberellin-response mutant. *Planta* **175**, 107–114.

Liang, X.W., Abel, S., Keller, J.A., Shen, N.F. and Theologis, A. (1992) The 1-aminocyclopropane-1-carboxylate synthase gene family of *Arabidopsis thaliana. Proc. Natl. Acad. Sci. USA* **89**, 11046–11050.

McCarty, D.R. (1992). The role of *VP1* in regulation of seed maturation in maize. *Biochem. Soc. Trans.* **20**, 89–92.

McCarty, D.R. (1995) Genetic control and integration of maturation and germination pathways in seed development. *Annu. Rev. Plant Physiol. Plant Mol. Biol.* **46**, 71–93.

McCarty, D.R., Carson, C.B., Stinard, P.S. and Robertson, D.S. (1989). Molecular analysis of *viviparous-1*: an abscisic acid insensitive mutant of maize. *Plant Cell* **1**, 523–532.

McCarty, D.R., Hattori, T., Carson, C.B., Vasil, V., Lazar, M. and Vasil, I.K. (1991) The *viviparous-1* developmental gene of maize encodes a novel transcriptional activator. *Cell* **66**, 895–905.

Manglesdorf, P.C. (1930) The inheritance of dormancy and premature germination in maize. *Genetics* **15**, 462–494.

Meinke, D.W., Franzman, L.H., Nickle, T.C. and Yeung, E.C. (1995) Leafy cotyledon mutants of *Arabidopsis. Plant Cell* **6**, 1049–1064.

Neill, S.J., Horgan, R. and Parry, A.D. (1986) The carotenoid and abscisic acid content of viviparous kernels and seedlings of *Zea mays*. L. *Planta* **169**, 87–96.

Neill, S.J., Horgan, R. and Rees, A.F. (1987) Seed development and vivipary in *Zea mays* L. *Planta* **171**, 358–364.

Olszewski, N.E. and Jacobsen, S.E. (1993) Mutations at the *SPINDLY* locus of *Arabidopsis* alter gibberellin signal transduction. *Plant Cell* **5**, 887–896.

Parcy, F., Valon, C., Raynal, M., Gaubiercomella, P., Delseny, M. and Giraudat, J. (1994) Regulation of gene-expression programs during *Arabidopsis* development: roles of the *Abi3* locus and of endogenous abscisic acid. *Plant Cell* **6**, 1567–1582.

Robertson, D.S. (1955) The genetics of vivipary in maize. *Genetics* **40**, 745–760.

Smith, J.D. and Neuffer, M.G. (1992) *Viviparous10*: a new viviparous mutant in maize. *Maize Genet. Coop. News* **66**, 34.

Vasil, V., Marcotte, W.R. Jr, Rosenkrans, L., Cocciolone, S.M., Vasil, I.K., Quatrano, R.S. and McCarty, D.R. (1995) Overlap of *Viviparous1* (*VP1*) and abscisic acid response elements in the Em promoter: G-box elements are sufficient, but not necessary for *VP1* transactivation. *Plant Cell* **7**, 1511–1518.

Walbot, V. (1992) Strategies for mutagenesis and gene cloning using transposon tagging and T-DNA insertional mutagenesis. *Annu. Rev. Plant Physiol. Plant Mol. Biol.* **43**, 49–82.

Novel embryo-specific barley genes: are they involved in desiccation tolerance?

D. Bartels, R. Roncarati and R. Alexander

9.1 Introduction

9.1.1 *The developmental programme of embryogenesis*

During embryogenesis a defined programme of events allows a zygote to develop into a seed. During this process the new plant is established and nutrient reserves for the germinating seedling accumulate. Embryogenesis terminates with a period of dormancy. According to Rogers and Quatrano (1983),embryogenesis in wheat — which is representative of monocotyledonous plants — can be divided into five stages based on morphological markers. Stage 1 is characterized by rapidly proliferating cells, resulting in an undifferentiated embryo, and stage 2 is characterized by the continuation of rapid cell divisions, leading to differentiated embryonic structures. Stage 3 is characterized by the completion of tissue differentiation, and stage 4 involves cell expansion which determines the final size of the embryo, and storage protein deposition. Stage 5 involves the onset of desiccation leading to developmental arrest. Different gene sets must become active during the different morphological stages, and their gene products are responsible for the correct course of development. Detailed analysis of embryogenesis in cotton has shown that the expression of these genes is organized in temporally defined modules (Galau *et al.* 1987; Hughes and Galau, 1989).

Genes expressed during the different stages of embryogenesis have been described from a large number of plant species. However, the regulation and function of most of these gene products is not understood. The mutational dissection of embryogenesis programmes is a very active field of research which has made major contributions to unravelling processes of embryogenesis. Of the monocotyledonous plants, a large number of mutants have been described for maize (Sheridan, 1988), but the largest source of information has been provided by *Arabidopsis* mutants with defects in embryo development (reviewed by Goldberg *et al.*,1994). This is a rapidly growing field which is described in more detail in other contributions to this book (see Chapters 2, 3, 4 and 6). The objective of this chapter is to focus on the acquisition of desiccation tolerance as one particular event of embryogenesis. During normal development, the seed loses up to 25% of its initial water content in the last stage of embryogenesis, and then remains in this dormant state, which is a requirement for later germination.

9.1.2 *Role of ABA in late embryogenesis*

Observations made with embryos cultured *in vitro*, measurements of hormone concentrations and analyses of mutants suggest that the plant hormone abscisic acid (ABA) plays an important regulatory role in late embryogenesis. When excised embryos are cultured *in vitro* in the presence of ABA, precocious germination is prevented and the typical embryogenic gene expression programme is maintained. Many examples of this have been demonstrated for embryos from both monocotyledonous and dicotyledonous plants (for example, see Finkelstein *et al.*, 1985; Quatrano, 1986). Precocious germination *in vivo* occurs in so-called viviparous mutants of maize and *Arabidopsis* (Koornneef *et al.*, 1989; Robichaud *et al.*, 1980). Some of these mutants are characterized by a reduced sensitivity to ABA, and two of their pleiotrophic phenotypes are confined to seed development. Two of the genes responsible for these phenotypes have been isolated: the *VP1* gene from maize which encodes a potential transcription activator (McCarty *et al.*, 1991) and the *ABI-3* gene from *Arabidopsis* which shares high sequence similarity with *VP1* (Giraudat *et al.*, 1992). The *ABI-3* gene product controls the expression of a set of seed-specific mRNAs which interact with ABA and other developmental factors (Parcy *et al.*, 1994). Another line of evidence which supports a regulatory role for ABA in gene expression is derived from promoter studies which have revealed promoter elements conferring ABA responsiveness to reporter genes (e.g. Marcotte *et al.*, 1989; Skriver *et al.*, 1991). However, there is some controversy, mainly derived from detailed analysis of gene expression in cotton embryogenesis, as to whether ABA has the dominant regulatory role in seed development. This work postulates a maternal maturation factor and a post-abscission factor as major regulatory components for embryogenesis (Galau *et al.*, 1991; Hughes and Galau, 1991).

9.1.3 *Gene expression during late embryogenesis*

Distinct mRNA sets are expressed during the different developmental stages of embryogenesis. A characteristic stage of late embryogenesis occurs when the seed loses water and expresses late embryogenesis-bundant (*LEA*) genes. A large number of such *LEA* genes have been isolated and characterized from a variety of different plants, and they have been grouped into classes based on sequence homologies. Although the function of these *LEA* genes has not been confirmed experimentally, the mode of their expression and their proposed biochemical properties have led to the suggestion that they have a role as protectants during desiccation (Baker *et al.*, 1988; Dure, 1993). The *LEA* genes appear to occur ubiquitously among plants, and structural sequence motifs are conserved within the classes of *LEA* genes across different plant species. Two of the best character-ized *LEA* genes are representatives of the *LEA1* and *LEA2* groups. The Em gene (Cuming and Lane, 1979) is one of the most extensively studied genes of the *LEA1* group, and it is discussed by Cuming *et al.* in Chapter 7 of this volume.

Briefly, studies on the ABA regulation of Em have revealed a putative ABA-responsive element (ABRE) in the Em promoter (Marcotte *et al.*, 1989), and a gene encoding a leucine zipper protein that bound specifically to this sequence has been cloned (Guiltinan *et al.*, 1990). Recently, three members of a small multigene family which show similarities to Em have been cloned from barley (Espelund *et al.*, 1992). The deduced amino acid sequences are conserved throughout the whole protein in all members of this group, which include the homologous genes from wheat, maize, radish, carrot and *Arabidopsis* (Gaubier *et al.*, 1993; Litts *et al.*, 1987; Raynal *et al.*, 1989; Ulrich *et al.*, 1990; Williams and Tsang, 1991). The proteins are characterized by a high level of hydrophilicity, and this hydrophilic component may be present once, as in Em, but can be present three or four times, as in the corresponding barley clones. The function of these hydrophilic regions may be to bind water and thereby generate a local aqueous environment that protects intracellular structures. Nevertheless, a more complicated and specialized mechanism has also been suggested, based on the different numbers of hydrophilic repeats and the differential regulation of the corresponding genes (Espelund *et al.*, 1992).

Group 2 *LEA* genes show sequence homologies with the *LEA* D11 gene from cotton (Baker *et al.*, 1988) which is representative of this group that includes a set of genes rapidly induced by ABA (RAB; Mundy and Chua, 1988) or water stress (dehydrins in barley and maize; Close *et al.*, 1989). These genes are expressed in both zygotic and somatic tissues of a variety of plants. Group 2 *LEA* proteins are hydrophilic, remain soluble after boiling, and show conservation of a lysine-rich motif (KIKEKLPG). The dehydrins have been shown to be distributed between the cytoplasm and the nucleus of various cell types in mature maize grains (Asghar *et al.*, 1994; Goday *et al.*, 1993).

9.2 The barley experimental system

The acquisition of desiccation tolerance is an obligatory programme of embryo-genesis in all orthodox seeds. Hence developing embryos are an excellent experimental system for the molecular analysis of components which contribute to desiccation tolerance. The genetic information for desiccation tolerance in seeds is under strict developmental control restricted to specific phases of embryo-genesis. In monocotyledonous plants the synthesis and deposition of reserves occurs mainly in the endosperm tissue, and for this reason monocots are preferred as experimental systems, since it is possible to separate the metabolic events of nutrient synthesis and acquisition of desiccation tolerance. In order to establish such an experimental system, developing barley embryos were isolated, desiccated and tested *in vitro* for their capacity to germinate precociously; this approach is illustrated in Figure 9.1. Dissected barley embryos are able to germinate precociously as early as 8 days after pollination (DAP). However, these embryos do not germinate after a severe dehydration treatment to < 10% of initial water content; they are thus desiccation intolerant. Nevertheless, at a precise later stage during embryogenesis the barley embryos acquire desiccation tolerance, and all

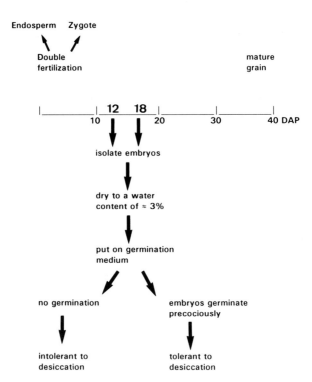

Development of a barley grain

Figure 9.1. *Diagram illustrating the experimental approach used to isolate desiccation-intolerant and desiccation tolerant barley embryos. The upper part depicts grain development, starting with fertilization and resulting in the production of mature barley grains. Along the time axis, expressed as days after pollination (DAP), embryos were isolated from the middle part of the ear. The aseptic embryos were either placed on germination medium or dried on filter paper for 16 h in a ventilating hood and then transferred to germination medium. For this experiment, barley plants of the cultivar Aura were grown in a Conviron growth chamber under controlled light and temperature conditions (for details see Bartels* et al., *1988).*

isolated embryos at 18 DAP germinate after drying to a water content as low as 3% (Bartels *et al.*, 1988). ABA treatment of desiccation-intolerant embryos leads to the induction of desiccation tolerance (Bartels *et al.*, 1988; Bocchiccio *et al.*, 1988). The transition from the desiccation-intolerant to the desiccation-tolerant stage was associated with a change in the expression pattern of mRNAs and proteins. A set of 25–30 gene products, which are absent or down-regulated in intolerant embryos (12 DAP), are induced after 16 DAP, namely at the time when desiccation tolerance is acquired.

9.3 Isolation of cDNA clones

In order to isolate transcripts which potentially contribute to desiccation tolerance in developing barley embryos, a cDNA clone bank was constructed from poly(A) RNA from desiccation-tolerant embryos (18 DAP). The cDNA clones were screened differentially using [^{32}P]-labelled RNA probes from 12 DAP desiccation-intolerant and 18 DAP desiccation-tolerant embryos. cDNA clones were selected which preferentially hybridized with the 18 DAP probe. The cDNA clones were classified into several main groups according to their expression patterns. The majority of the cDNA clones encoded transcripts that were expressed not only in embryos but also in other tissues (leaves and roots); these clones were discarded. For the investigation of desiccation tolerance in embryos, a small group of clones was selected. The transcripts encoded by these particular clones are embryo specific and their expression is temporarily correlated with the onset of desiccation tolerance. The abundance of selected transcripts during embryogenesis is shown in Figure 9.2, and some features of these cDNA clones are summarized in Table 9.1. The cDNA clones pG 22-69, pG 31 and pG 38-83 had not been entered into available databanks before, whereas pG 30-44 was revealed to be identical with the wheat Em transcript as described by Espelund *et al.* (1992). For comparison, the rice clone rab 16 (Mundy and Chua, 1988) was used in the RNA blot experiment as it is one of the most widely studied late-embryogenesis-abundant transcripts. The transcript level of pG 31 is relatively abundant in young embryos and does not change appreciably during embryogenesis, whereas pG 22-69 and pG 38-83 were selected for further analysis because higher transcript levels are correlated with the acquisition of desiccation tolerance. The relative abundance of these two transcripts resembles that of the Em homologue pG 30-44 and rab 16, both of which are postulated to be involved in processes related to desiccation tolerance in embryos, although their exact functions are still unclear.

9.4 pG 31 encodes a protein with homologies to bacterial glucose or ribitol dehydrogenase

The protein encoded by pG 31 comprises an open reading frame of 293 amino acids. As is shown in Figure 9.3, this amino acid sequence revealed significant homologies (30% amino acid identity) to the bacterial enzymes glucose dehydrogenase (GDH: Heilmann *et al.*, 1988; Jany *et al.*, 1984) and ribitol dehydrogenase

Table 9.1. *Barley embryo-specific transcripts*

cDNA clone	Protein encoded	Reference
pG 22-69	Aldose-reductase homologue	Bartels *et al.* (1991)
pG 31	Bacterial glucose dehydrogenase Ribitol dehydrogenase	Alexander *et al.* (1994)
pG 38-83	?	Alexander (1992), and Bartels, unpublished data
pG 30-44	Em of wheat, B1.19 of barley	Bartels and Velasco, unpublished data, and Espelund *et al.* (1992)

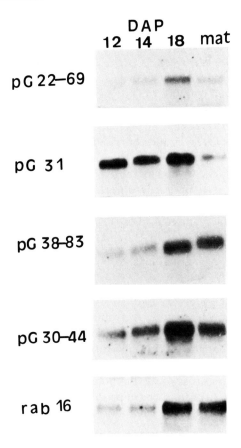

Figure 9.2. *Northern blot analysis with selected cDNA clones. Poly A+ RNAs were isolated from developing barley embryos at 12, 14 and 18 DAP as well as from mature embryos. Two µg of RNA were separated electrophoretically on a 1.2% (w/v) formaldehyde agarose gel using a formaldehyde N-morpholino-propanesulphonic acid (MOPS) buffer system. Several identical nylon filters carrying these RNAs were prepared and hybridized with [³²P]-labelled inserts of the indicated cDNAs. Details of the cDNA clones are given in Table 9.1.*

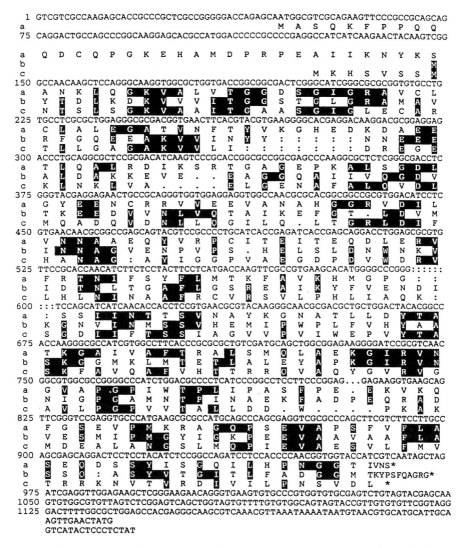

Figure 9.3. *Comparison of the amino acid sequences deduced from (a) the barley cDNA clone pG 31, (b) glucose dehydrogenase of* Bacillus megaterium *(Heilmann* et al.*, 1988) and (c) ribitol dehydrogenase from* Klebsiella aerogenes *(Loviny* et al.*, 1985). Identical amino acids are printed in black; gaps were introduced to optimize matches. The upper line shows the nucleotide sequence of the cDNA clone pG 31. This figure was reproduced from Alexander* et al. *(1994) with the permission of Springer-Verlag, Berlin.*

(RDH: Loviny *et al.*, 1985). Ribitol dehydrogenase catalyses the oxidation of ribitol to ribose, and glucose dehydrogenase oxidizes D-glucose to gluconic acid with NAD$^+$ or NADP$^+$ as coenzyme (Figure 9.4b). The reactions catalysed by these bacterial enzymes are substrate specific, and other sugars such as mannose, galactose, ribose, fructose or myo-inositol cannot be utilized (Heilmann *et al.*, 1988). The two bacterial dehydrogenases belong to a family of polyoldehydro-genases which have previously been described from bacteria, yeast, *Drosophila* and mammals, but not from plants (Jörnvall *et al.*, 1984). The pG 31 cDNA clone represents the first member of this family to be isolated from plants (Alexander *et al.*, 1994). The demonstration of a plant gene belonging to the family of glucose dehydrogenases and ribitol dehydrogenases extends the previously reported evolutionary scheme (Jörnvall *et al.*, 1984) to the plant kingdom. The homologies between the barley pG 31 protein and glucose or ribitol dehydrogenases are across the complete protein sequences. Three glycine residues in the amino-terminal part of the glucose dehydrogenase molecule are functionally important and are involved in binding the coenzymes NAD$^+$ or NADP$^+$ (Heilmann *et al.*, 1988); these glycine residues (Gly-47, Gly-51 and Gly-53) are conserved in the barley pG 31 protein. In bacteria, the active glucose dehydrogenase is a homo-tetramer; the tyrosine residue at position 200, involved in the association of the subunits, is also conserved in the barley protein. The observation that functionally important amino acids have been conserved among plants and bacteria suggests that metabolic pathways similar to those found in bacteria must also exist in plants. This hypothesis was tested by assaying crude protein preparations for sugar dehydrogenase activities. In contrast to general carbohydrate metabolism, which mainly involves phosphoryl-ated sugar molecules, the glucose dehydrogenase described here catalyses the oxidation of unphosphorylated glucose. To test whether such enzymatic activities can be related to the presence of the pG 31 transcript, protein fractions from different tissues were assayed for glucose dehydrogenase activity, which was measured spectrophotometrically following the formation of NAD(P)H in protein extracts containing 50 μmol D-glucose or ribitol. Enzymatic activity was detected in isolated embryos but not in germinated embryos or in leaves, and the enzymatic activity was specific for glucose as the substrate (Alexander *et al.*, 1994). It was demonstrated that this enzymatic activity was related to the pG 31 protein by conducting an experiment in which pG 31 polyclonal antiserum was included in the enzyme assays. In this experiment glucose dehydrogenase activity was substantially reduced, and no activity was detected after the protein–antigen complex had been removed with Protein A Sepharose. Enzyme assays with different sugars (glucose, sorbitol, xylulose, fructose and ribitol) showed that only D-glucose was specifically oxidized in embryo protein extracts. In the presence of the other sugars, some reducing activities were also observed, but they were not confined to embryo protein extracts alone. The results of these enzymatic assays point to a biochemical reaction novel for embryos, involving non-phosphorylated sugars (particularly D-glucose). The only criticism of the enzymatic data is that it is based on crude protein fractions. In order to substantiate the biochemical properties of the potential glucose dehydrogenase and the mechanism of its reaction, purification of the enzyme is necessary.

9.5 A barley gene with features of an aldose reductase

As well as the cDNA clone discussed above, a second cDNA clone, pG 22-69, derived from the differential screen of an 18 DAP embryo library, was associated with an enzyme of carbohydrate metabolism. pG 22-69 was shown to encode a barley homologue of an aldose or aldehyde reductase (Bartels *et al.*, 1991). Aldose reductase (EC1.1.1.21) and aldehyde reductase (E.C.1.1.1.2) are monomeric, primarily NADPH-dependent, oxido-reductases with a broad and overlapping substrate specificity ranging from aldose sugars to aromatic aldehydes (Flynn, 1982). Proteins related to this group are widely distributed in mammalian tissues, birds, amphibians, plants, fungi and bacteria (Grimshaw, 1992, and references therein). This suggests the existence of a highly evolved class of enzymes, although their exact physiological function is not clear in most examples. It is notable that, despite the evolutionary conservation, they possess inefficient kinetic properties. Aldose reductase has received particular attention due to its possible role in the reduction of glucose during diabetic hyperglycaemia. In this condition aldose reductase reduces glucose to sorbitol, which is then oxidized to fructose by sorbitol dehydrogenase (Bhatnagar and Srivastava, 1992). Under non-pathogenic conditions this pathway plays a minor role in glucose metabolism in most animal tissues. Another series of studies, mainly in renal tissues, points to an osmoregulatory function of aldose reductase, where the changes in intracellular sorbitol concentration help to maintain the osmotic balance (Bedford *et al.*, 1987). Aldose reductase mRNA and enzymatic activity have been found to increase in renal cells in response to an elevation of NaCl concentration in the medium and a high extracellular osmolarity (Bagnasco *et al.*, 1987). Interestingly, the homologies of the aldose and aldehyde reductase families extend to other proteins with enzymatic or structural functions, such as prostaglandin F synthase, 2,5-diketogluconic acid reductase from *Corynebacterium* and rho-crystallin from frog. They have been classified as members of the aldo-keto-reductase superfamily (Bartels *et al.*, 1991; Bohren *et al.*, 1989).

The question under consideration in the present chapter is the significance of a barley transcript belonging to the aldose reductase family. As the function of the barley gene is not yet confirmed, it will be referred to as the aldose reductase homologue (AR-h). The AR-h transcript belongs to a small set of genes which are only expressed in embryos, and which are induced at the onset of desiccation when the barley embryo is able to germinate despite a severe dehydration treatment. The level of the transcript is developmentally regulated and can be modulated by exogenous ABA treatments (Bartels *et al.*, 1991).

In order to understand the role of the AR-h transcript, a functional analysis of the encoded protein was required. Indications that the barley AR-h protein could possess aldose reductase activity came from the observation that ammonium sulphate fractions of protein extracts from 18 DAP barley embryos exhibited enzymatic aldose reductase activity. Antibodies produced against the barley AR-h protein synthesized *in vitro* detected a 34-kDa protein specifically in protein fractions of 18 DAP embryos (Bartels *et al.*, 1991). To investigate the enzymatic properties of the barley protein in more detail, two experimental approaches were used: the purification of the native protein from 18 DAP barley embryos, and the synthesis of the corresponding recombinant protein together with the synthesis of a mutant protein. Although a barley AR-h protein was purified from 18 DAP embryos using immunochromatography, which migrated as a single band of 34 kDa in an SDS-protein gel, the yields were generally low and did not allow further biochemical characterization (Roncarati et al., 1995). For this reason, most of the protein characterization has been performed using a purified recombinant protein.

Among the aldose reductases, the tetrapeptide IPKS in the C-terminal part of the protein is conserved, and it has been suggested that the lysine residue participates in the binding to the coenzyme NADPH, or that it affects the catalytic activity (Bohren et al., 1991; Morjana et al., 1989). To test the importance of the lysine residue in the barley protein, a mutant recombinant protein was generated in which lysine was substituted by methionine (this will be referred to as the lys-mutant). Methionine was selected because it is almost isosteric with lysine, but lacks the positive charge of the ϵ-amino group which is probably involved in the formation of the bond with the coenzyme. The substrate specificity of both the wild-type and the mutant recombinant was tested using a series of aldo-sugars with different numbers of carbon atoms (C_3 to C_6). Only DL-glyceraldehyde and D-erythrose were significantly reduced in the presence of NADPH as coenzyme. Very low activities were found with xylose, glucose and galactose at concentrations of 400 mM (Roncarati et al., 1995). Phosphorylated derivatives of the sugars mentioned above performed even less efficiently. Aldoses with short carbon chains appear to be better utilized by this enzyme. The enzyme could not use NADH as cofactor, nor could it catalyse the reverse oxidation of sorbitol to glucose (Roncarati et al., 1995). The kinetic constants (K_m) of the wild-type and mutant enzyme were determined for the two substrates that were significantly reduced by the wild-type enzyme (DL-glyceraldehyde and erythrose), and for the coenzyme NADPH (Table 9.2). The wild-type enzyme exhibited similar K_m values for glyceraldehyde and erythrose ($K_{mGA} = 56$; $K_{mEry} = 45$), indicating that these substrates are used with approximately the same efficiency. The K_m value of the purified native barley protein was comparable with that of the wild-type recombinant protein. From the comparison of K_m values of wild-type and mutant proteins, it appears that the replacement of the ϵ-amino group of lysine causes a decrease in substrate affinity ranging from three- to sevenfold ($K_{mGA} = 159 \pm 28$; $K_{mEry} > 324$); the K_m for NADPH is also moderately increased (by about fivefold; Table 9.2). In all cases the mutant enzyme performed less efficiently than the wild-type enzyme. The lower catalytic activity of the mutant can be confirmed by comparing the specific activity of wild-type and mutant enzymes; the specific activity of the mutant is about fivefold lower than that of the wild-type aldose reductase. These results provide further support for the possible involvement of Lys-262 in NADPH binding. However, the K_m values of the wild type for glyceraldehyde and erythrose are three orders of magnitude higher than the corresponding values for animal enzymes (Bohren et al., 1991;

Table 9.2. *Apparent kinetic constants of recombinant aldose reductase proteins*

	Wild type (mM)	Mutant (Lys→Met) (mM)
K_m (GA)	56 ± 9	159 ± 28
K_m (erythrose)	45 ± 4	324 ± 50
K_m (NADPH)	0.2 ± 0.1	1.4 ± 0.2

Grimshaw, 1992). Extremely low activity was obtained with glucose, which should be the natural substrate for aldose reductase in the polyol pathway (Figure 9.4). However, the problem of physiological significance is intrinsic to all aldose reductases.

9.6 Distribution of glucose dehydrogenase and aldose reductase in the plant kingdom

Both barley cDNA clones, pG 31 encoding glucose dehydrogenase and pG 22-69 encoding aldose reductase, were the first of their class to be described for plants. The expression of both transcripts is embryo specific and abundant during certain stages of embryogenesis (Figure 9.2). Both transcripts code for proteins with the potential enzymatic function of catalysing reactions of carbohydrate metabolism using non-phosphorylated sugars or sugar derivatives as substrates. Furthermore, the encoded proteins of both genes belong to the large class of oxido-reductases. To date, no other plant genes with homologies to the glucose dehydrogenase class

a The Polyol Pathway

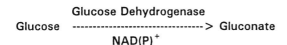

b The Glucose Dehydrogenase Reaction

Glucose -------------------------------> Gluconate
          Glucose Dehydrogenase
          NAD(P)$^+$

Figure 9.4. *Enzymatic reactions by (a) aldose reductase and (b) glucose dehydrogenase.*

have been described in the literature. Genes with a high level of homology to the barley aldose reductase have been isolated from embryogenic cultures of bromegrass (Lee and Chen, 1993), and a less closely related gene encoding a reductase co-acting with chalcone synthase has also been isolated from soybean (Welle *et al.*, 1991).

Both discoveries, the glucose dehydrogenase and the aldose reductase, support the view that functionally important DNA sequences appear to be conserved among plants, bacteria and some animals, and that similar metabolic pathways are expected to be active in all types of organism.

9.7 Conclusions

Both the glucose dehydrogenase and the aldose reductase transcripts are expressed at relatively high levels in embryos (Figure 9.2). Most other embryo-specific transcripts which have been classified as belonging to *LEA* genes are likely to have a structural function and are not active enzymes. Despite the demonstration of glucose dehydrogenase activity and aldose reductase activity, it must be emphasized that the apparent kinetic constants are high and indicate low substrate affinities for both enzymatic reactions using the obvious substrates in the assays. However, it could be argued that the physiologically 'correct' substrates have not yet been identified, and this is still partly an open question in the case of the human and animal aldose reductases. An interesting proposal is that these enzymes could function in cellular detoxification processes, rather than synthesizing osmolytes (Grimshaw, 1992). On the basis of kinetic parameters, Grimshaw has suggested that aldose reductase could be involved in the detoxification of a broad range of substrates which might otherwise lead to the formation of products harmful to the cell. Such a function could also be important in plants during desiccation.

On the basis of the research described here, tools have been made available to test the importance of the glucose dehydrogenase and aldose reductase genes from barley with regard to the acquisition of desiccation tolerance. A further experimental approach will involve the use of reverse genetics to create mutants. Barley plants can be transformed with antisense constructs for aldose reductase or glucose dehydrogenase under the control of a barley embryo-specific promoter, such as the one from the aldose reductase gene which has recently been isolated and characterized (Roncarati *et al.*, 1995).

Acknowledgements

We thank Prof. Salamini for guidance and advice during this work, M. Pasemann for typing the manuscript, Prof. J. Mundy for the Rab 17 cDNA clone from rice for Northern experiments, and Dr J. Ingram for linguistic corrections. Parts of the work described here were supported by a grant to D.B. from the European Community (Contract No. TS2-0030-D) and a European Community Fellowship to R.R. in the Biotechnology Programme.

References

Alexander, R. (1992) Untersuchungen zur Genexpression in Gerstenembryonen während der Ausbildung der Trockentoleranz. Ph.D. Thesis, University of Köln, Germany.

Alexander, R., Alamillo, J.M., Salamini, F. and Bartels, D. (1994) A novel embryo-specific barley cDNA clone encodes a protein with homologies to bacterial glucose and ribitol dehydrogenase. *Planta* **192**, 519–525.

Asghar R, Fenton RD, DeMason DA, Close TJ. (1994) Nuclear and cytoplasmic localization of maize embryo and aleurone dehydrin. *Protoplasma* **177**, 87–94.

Bagnasco, S.M., Uchida, S., Balaban, R.S., Kador, P.F. and Burg, M.B. (1987) Induction of aldose reductase and sorbitol in renal medullary cells by elevated extracellular NaCl. *Proc. Natl Acad. Sci. USA* **84**, 1718–1720.

Baker, J., Steele. C. and Dure, L. III (1988) Sequence and characterization of 6 Lea proteins and their genes from cotton. *Plant Mol. Biol.* **11**, 277–291.

Bartels, D., Singh, M. and Salamini, F. (1988) Onset of desiccation tolerance during development of the barley embryo. *Planta* **175**, 485–492.

Bartels, D., Engelhardt, K., Roncarati, R., Schneider, K., Rotter, M. and Salamini, F. (1991) An ABA and GA modulated gene expressed in the barley embryo encodes an aldose reductase related protein. *EMBO J.* **10**, 1037–1043.

Bedford, J.J., Bagnasco, S.M., Kador, P.F., Harris, H.W. Jr and Burg, M.B. (1987) Characterization and purification of a mammalian osmoregulatory protein, aldose reductase, induced in renal medullary cells by high extracellular NaCl. *J. Biol. Chem.* **262**, 14255–14259.

Bhatnagar, A. and Srivastava, S.K. (1992) Aldose reductase: congenial and injurious profiles of an enigmatic enzyme. *Biochem. Med. Metab. Biol.* **48**, 91–121.

Bocchiccio, A., Vazzana, C., Raschi, A., Bartels, D. and Salamini, F. (1988) Effect of desiccation on isolated embryos of maize. Onset of desiccation tolerance during development. *Agronomie* **8**, 29–36.

Bohren, K.M., Bullock, B., Wermuth, B. and Gabbay, K.H. (1989) The aldo-keto reductase superfamily. *J. Biol. Chem.* **264**, 9547–9551.

Bohren, K.M., Page, J.L., Shankar, R., Henry, S.P. and Gabbay, K.H. (1991) Expression of human aldose and aldehyde reductases. *J. Biol. Chem.* **266**, 24031–24037.

Close, T.J., Kortt, A.A. and Chandler, P.M. (1989) A cDNA-based comparison of dehydration-induced proteins (dehydrins) in barley and corn. *Plant Mol. Biol.* **13**, 95–108.

Cuming, A.C. and Lane, B.G. (1979) Protein synthesis in imbibing wheat embryos. *Eur. J. Biochem.* **145**, 351–357.

Dure, L. III (1993) A repeating 11-mer amino acid motif and plant desiccation. *Plant J.* **3**, 363–369.

Espelund, M., Sæbøe-Larssen, S., Hughes, W.D., Galau, G.A., Larsen, F. and Jacobsen, K.S. (1992) Late embryogenesis-abundant genes encoding proteins with different numbers of hydrophilic repeats are regulated differentially by abscisic acid and osmotic stress. *Plant J.* **2**, 241–252.

Finkelstein, R.R., Tembarge, K.M., Shumway, J.E. and Crouch, M.L. (1985) Role of ABA in maturation of rape seed embryos. *Plant Physiol.* **78**, 630–636.

Flynn, T.G. (1982) Aldehyde reductases. Monomeric oxidoreductases with multifunctional potential. *Biochem Pharmacol.* **31**, 2705–2712.

Galau, G.A., Bijaisoradat, N. and Hughes, D.W. (1987) Accumulation kinetics of cotton late embryogenesis-abundant mRNAs and storage protein mRNAs: coordinate regulation during embryogenesis and the role of abscisic acid. *Dev. Biol.* **123**, 198–212.

Galau, G.A., Jakobsen, K.S. and Hughes, D.W. (1991) The control of late dicot embryogenesis and early germination. *Physiol. Plant.* **81**, 280–288.

Gaubier, P., Raynal, M., Hull, G., Huestis, G.M., Grellet, F., Arenas, C., Pagès, M. and Delseney, M. (1993) Two different *Em*-like genes are expressed in *Arabidopsis thaliana* seeds during maturation. *Mol. Gen. Genet.* **238**, 409–418.

Giraudat, J., Hauge, B.M., Valon, C., Smalle, J., Parcy, F. and Goodman, H.M. (1992) Isolation of the *Arabidopsis ABI3* gene by positional cloning. *Plant Cell* **4**, 1251–1261.

Goday, A., Jensen, A.B., Culiánez-Macià, F.A., Mar Albà, M., Figueras, M., Serratosa, J., Torrent, M. and Pagès, M. (1993) The maize abscisic acid-responsive protein Rab17 is located in the nucleus and interacts with nuclear localization signals. *Plant Cell* **6**, 351–360.

Goldberg, R.B., de Paiva, G. and Yadegari, R. (1994) Plant embryogenesis: zygote to seed. *Science* **266**, 605–614.

Grimshaw, C.E. (1992) Aldose reductase: model for a new paradigm of enzymic perfection in detoxification catalysts. *Biochemistry* **31**, 10139–10145.

Guiltinan, M.J., Marcotte, W.R. and Quatrano, R.S. (1990) A plant leucine zipper protein that recognizes an abscisic acid response element. *Science* **250**, 267–270.

Heilmann, H.J., Mägert, H.J. and Gassen, H.G. (1988) Identification and isolation of glucose dehydrogenase genes of *Bacillus megaterium* M1286 and their expression in *Escherichia coli*. *Eur. J. Biochem.* **174**, 485–490.

Hughes, D.W. and Galau, G.A. (1989) Temporally modular gene expression during cotyledon development. *Genes Dev.* **3**, 358–369.

Hughes, D.W. and Galau, G.A. (1991) Developmental and environmental induction of Lea and LeaA mRNAs and the post-abscission program during embryo culture. *Plant Cell* **3**, 605–618.

Jany, K.D., Ulmer, W., Fröschle, M. and Pfleiderer, G. (1984) Complete amino acid sequence of glucose dehydrogenase from *Bacillus megaterium*. *FEBS Lett.* **165**, 6–10.

Jörnvall, H., Von Gahr-Lindström, H., Jany. K,D,, Ulmer, W. and Fröschle, M. (1984) Extended superfamily of short alcohol-polyol-sugar dehydrogenases: structural similarities between glucose and ribitol dehydrogenases. *FEBS Lett.* **165**, 190–196.

Koornneef, M., Hanhart, C.J., Holhorst, H.W.M. and Karssen, C.M. (1989) *In vivo* inhibition of seed development and reserve protein accumulation in recombinants of abscisic acid biosynthesis and responsiveness mutants in *Arabidopsis thaliana*. *Plant Physiol.* **90**, 463–469.

Lee, S.P. and Chen, T.H.H. (1993) Molecular cloning of abscisic acid-responsive mRNAs expressed during the induction of freezing tolerance in bromegrass (*Bromus inermis* Leyss) suspension culture. *Plant Physiol.* **101**, 1089–1096.

Litts, J.C., Colwell, G.W., Chakerian, R.L. and Quatrano, R.S. (1987) The nucleotide sequence of a cDNA clone encoding the wheat Em protein. *Nucleic Acids Res.* **15**, 3607–3618.

Loviny, R., Norton, P.M. and Hartley, B.S. (1985) Ribitol dehydrogenase of *Klebsiella aerogenes*. *Biochem. J.* **230**, 579–585.

Marcotte, W.R., Russell, S.H. and Quatrano, R.S. (1989) Abscisic acid-responsive sequence from the Em gene of wheat. *Plant Cell* **1**, 969–976.

McCarty, D.R., Hattori, T., Carson, C.B., Vasil, V., Lazar, M. and Vasil, I.K. (1991) The *Viviparous 1* developmental gene of maize encodes a novel transcriptional activator. *Cell* **66**, 895–905.

Morjana, N.A., Lyons, C. and Flynn, T.G. (1989) Aldose reductase from human psoas muscle:

affinity labeling of an active site lysine by pyridoxal 5'-phosphate and pyridoxal 5'-diphospho-5' adenosine. *J. Biol. Chem.* **264**, 2912–2919.

Mundy, J. and Chua, N.-H. (1988) Abscisic acid and water-stress induce the expression of a novel rice gene. *EMBO J.* **7**, 2279–2286.

Parcy, F., Valon, C., Raynal, M., Gaubier-Comella, P., Delseny, M. and Giraudat, J. (1994) Regulation of gene expression programs during *Arabidopsis* seed development: roles of the *ABI3* locus and of endogenous abscisic acid. *Plant Cell* **6**, 1567–1582.

Quatrano, R.S. (1986) Regulation of gene expression by abscisic acid during angiosperm development. *Oxford Surv. Plant Mol. Cell Biol.* **3**, 457–477.

Raynal, M., Depigny, D., Cooke, R. and Delseny, M. (1989) Characterization of a radish nuclear gene expressed during late seed maturation. *Plant Physiol.* **91**, 829–836.

Robichaud, C.S., Wong, J. and Sussex, I.M. (1980) Control of *in vitro* growth of viviparous embryo mutants of maize by abscisic acid. *Dev Genet.* **1**, 325–330.

Rogers, S.O. and Quatrano, R.S. (1983) Morphological staging of wheat caryopsis development. *Am. J. Bot.* **70**, 308–311.

Roncarati, R., Salamini, F. and Bartels, D. (1995) An aldose reductase homologous gene from barley: regulation and function. *Plant J.* **7**, 809–822.

Sheridan, W.F. (1988) Maize developmental genetics: genes of morphogenesis. *Annu. Rev. Genet.* **22**, 353–385.

Skriver, K., Olsen, F.L., Rogers, J.C. and Mundy, J. (1991) *Cis* acting elements responsive to gibberellin and its antagonist abscisic acid. *Proc. Natl Acad. Sci. USA* **88**, 7266–7270.

Ulrich, T.U., Wurtele, E.S. and Nikolau, B.J. (1990) Sequence of EMB-1, an mRNA accumulating specifically in embryos of carrot. *Nucleic Acids Res.* **18**, 2826.

Welle, R., Schröder, G., Schlitz, E., Griseback, H. and Schröder, J. (1991) Induced plant responses to pathogen attack. Analysis and heterologous expression of the key enzyme in the biosynthesis of phytoalexins in soybean (*Glycine max* L. Merr. cv. Harosoy 63). *Eur. J. Biochem.* **196**, 423–430.

Williams, B. and Tsang, A. (1991) A maize gene expressed during embryogenesis is abscisic acid-inducible and highly conserved. *Plant Mol. Biol.* **16**, 919–923.

The reprogrammed embryo: the endosperm as a quick route to understanding embryogenesis?

R.C. Brown, B. Lemmon, D. Doan, C. Linnestad and O.-A. Olsen

10.1 Introduction

An evolutionary event of paramount importance was the introduction of the endosperm into the reproductive cycle of flowering plants. Not only is endosperm the principal foodstuff upon which civilization depends, but as a developmental system it is of great interest to biologists. Although the endosperm and zygote have a common origin as products of double fertilization in angiosperms, their developmental pathways are very different. Regardless of the phylogenetic origin of the endosperm, it is ontogenetically the result of a fertilization event and therefore can be viewed as a reprogrammed embryo. This chapter summarizes the unusual developmental programme of endosperm, which includes an early syncytial stage and four types of wall deposition prior to the introduction of typical plant-like development. We emphasize the value of endosperm as a tractable system for investigating the fundamental controls of plant development, including those which are likely to play a central role in embryogenesis.

10.2 The evolutionary history of endosperm

Endosperm may be regarded as a third, short-lived generation in the life cycle of flowering plants. The function of the endosperm is to provide nourishment for development of the embryo that begins the new sporophyte generation (Gifford and Foster, 1989). Direct evidence in support of this function has recently become available. A chitinase has been isolated from the seeds of carrot which is present only in the endosperm. When present in culture (conditioned) medium, this chitinase can rescue developing mutant embryos (see Chapter 11, Toonen and de Vries).

The evolution of this altruistic third generation has long baffled plant scientists (for interesting contemporary reviews see Battaglia, 1980; Favre-Ducharte, 1984; Friedman, 1994). The endosperm may have originated as a continuation of gametophytic development triggered by the second fertilization event or, alternatively, as a supernumerary embryo reprogrammed to play a nutritive role. Recent evidence (reviewed by Friedman, 1994) supports the view that the endosperm originated from supernumerary embryos occurring in the immediate ancestors of the angiosperms. According to this interpretation, the endosperm is a highly modified evolutionary homologue of the embryo (Friedman, 1994). Although double fertilization occurs in gymnosperms, it does not result in the development of endosperm (Friedman, 1992). The nutritive tissue of gymnosperms is produced earlier as a massive female gametophyte tissue, and the second fertilization results in an additional embryo that competes for resources during development. In species of the gymnosperm *Ephedra*, two sperms from a single pollen tube regularly fertilize the egg nucleus and the sister ventral canal nucleus within an archegonium. The twin zygotes both develop into proliferating embryos. In angiosperms, the second fertilization results in a primary endosperm nucleus that enters a developmental pathway different from that of an embryo. Interestingly, both the gametophyte and the embryos of gymnosperms undergo a coenocytic stage before they become cellular. Although free nuclear embryo development is extremely rare in flowering plants, it has been reported in the genus *Paeonia* (reviewed by Yakovlev, 1967).

The two products of fertilization in angiosperms are closely related genetically; the egg and one of the polar nuclei are sister cells (Huang and Russell, 1992), and the two sperm result from a mitotic division of the generative cell. The level of ploidy in endosperm may vary depending upon the number of polar nuclei that unite with a sperm to give rise to the primary endosperm nucleus. In *Oenothera*, for example, a single polar nucleus is fertilized and the primary endosperm nucleus is diploid (Favre-Ducharte, 1984). In cereals, two polar nuclei unite with a sperm, and the resulting cell is triploid. Nevertheless, the maternal genomes may be assumed to be identical, since all cells of the embryo sac result from mitoses of a single megaspore.

10.3 Features of the endosperm system

As a system for studying development in higher plants, endosperm is without peer. It is abundant and matures rapidly to form a simple system consisting of only two principal tissues, aleurone and starchy endosperm. In barley the two tissues can be separated mechanically (Jakobsen *et al.*, 1989). The embryo and endosperm, which are closely related genetically and have a common starting point at double fertilization, develop simultaneously within the ovule. This situation allows direct comparison of the response of the various tissues of the parental sporophyte, embryo and endosperm to experimentation, and provides the opportunity for analysis of genetic controls of the essential processes of development. Importantly, mutants of the endosperm can be maintained as heterozygotes (in a fashion similar to the use of conditional mutants in other genetic systems), allowing the analysis of homozygous mutants expressed in endosperm that would be lethal if expressed in the sporophyte embryo. Numerous genetic mutants which affect endosperm development are known (Bosnes *et al.*, 1987, 1992; Felker *et al.*, 1985; Neuffer and Sheridan, 1980; Sheridan and Neuffer, 1980). Mutants that block various stages of endosperm development may indicate check points in the developmental pathway and may help to decipher the genetic mechanisms that direct morphogenesis.

Endosperm development is especially remarkable in that it begins as a coenocyte and culminates with the typical plant-like development of the peripheral aleurone (Brown *et al.*, 1994; Olsen *et al.*, 1995). In the process, endosperm displays an unparalleled diversity in the modes of wall deposition. The precise regulation of cell division pattern and subsequent expansion and shaping of cells are fundamental processes in the development of form in plants. Our data suggest that endosperm can serve as a model system in which the basic mechanisms of plant development, such as polarity, patterning, cell division and wall development, can be investigated. This complex process is thought to be controlled by the sequentially regulated expression of a genetic programme (Olsen *et al.*, 1992; Sheridan, 1988). Deciphering the regulation and expression of the genetic programme which controls the precisely oriented and complex processes of wall deposition is central to the understanding of plant development.

10.4 Developmental map of barley endosperm

A model for the development of barley endosperm, based on information from structural and genetic studies (Bosnes *et al.*, 1992; Olsen *et al.*, 1992), proposed that development occurs in four major stages: syncytial, cellularization, differentiation and maturation (Figure 10.1). Our recent study, using techniques of indirect immunofluorescence in sectioned material of carefully staged developing endosperm of barley (Brown *et al.*, 1994), provided data on the relationship between microtubules and the first three stages of endosperm morphogenesis (Figure 10.2). Such data permit analysis of the organization and function of the cytoskeleton associated with patterns of cellularization.

During the 21-day period during which the two barley endosperm tissues (starchy and aleurone) develop to maturity, four developmentally distinct types of wall are produced sequentially. Each type is associated with a different pattern of microtubules. During the syncytial stage before walls are deposited, nuclear-based radial systems of microtubules define domains of cytoplasm around nuclei and determine the pattern of cellularization. During a hiatus in mitotic activity between 5 and 7 days after pollination (DAP), the nuclear-cytoplasmic domains (NCDs) become polarized and elongate in an axis perpendicular to the embryo sac wall. The NCDs at this stage (arboreal) resemble trees when viewed from the side, with elliptical nuclei in the trunks and microtubules extending into

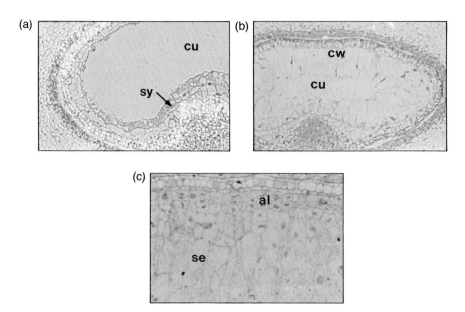

Figure 10.1. *Representative stages of developing wild-type barley endosperm as illustrated by light micrographs of transverse sections. (a) Syncytial endosperm (sy) surrounding the central vacuole (cv) of the central cell. (b) Endosperm cell walls (cw) grow in towards the centre of the central vacuole. (c) The endosperm consists of peripheral layers of aleurone cells (al) surrounding the starchy endosperm.*

root-like extensions at the embryo sac wall and into a canopy of cytoplasm adjacent to the central vacuole. The first anticlinal walls form in the absence of phragmoplasts between the NCDs, and compartmentalize the syncytium into open-ended alveoli. At about 7 DAP there is a sudden onset of phragmoplast development adventitiously in the cytoplasm where opposing sets of microtubules from adjacent alveoli interact and the second type of wall development is introduced. Thereafter, the anticlinal walls are guided by the adventitious phragmoplasts. Mitosis resumes in the alveoli, moving in a wave from the ventral to the dorsal surface. These mitoses are followed by typical interzonal phragmoplast/cell plates that result in periclinal walls. The daughter nuclei arising from this critical division have different developmental fates; those to the interior give rise to the starchy endosperm, while the peripheral cells form aleurone. With the resumption of mitotic activity in the peripheral aleurone layer, the predictive component of the cytokinetic apparatus, the preprophase band of microtubules (PPB), is introduced into the microtubule cycle. The microtubule cycle during development of the three-layered aleurone is entirely typical of plant histogenesis (Figure 10.2). Genes isolated by differential screening correspond in temporal and spatial pattern of expression with the stages of cellularization (See Section 10.7).

10.5 Mutants affecting endosperm development

Plants segregating for failure of the grain to fill properly have frequently been observed in maize and barley. The phenotypes of such mutant grains are referred to as *dek* (*defective*) or *shr* (*shrunken*) in maize (Clark and Sheridan, 1986; Lowe and Nelson, 1946; Mangelsdorf, 1926; Neuffer and Sheridan, 1980; Sheridan and Neuffer, 1980), and as *dex* or *sex* (*defective or shrunken endosperm expressing xenia*: segregation of mutant and normal endosperm phenotypes in heterozygous plants) in barley (Bosnes *et al.*, 1987, 1992; Ramage and Crandall, 1981). Our collection of barley *sex* mutants currently includes more than 100 entries, approximately half of which have been partially characterized by light and/or electron microscopy.

Based solely on the phenotype of endosperm, barley mutants fall into two main groups, referred to as Types I and II (Bosnes *et al.*, 1992). Type I mutants exhibit arrest at one of the developmental stages recognized in wild-type endosperm. Examples of mutants which display Type I developmental arrest include B7, in which development is arrested at the syncytial stage prior to vacuolation, B9, which lacks aleurone cells, N2, in which the anticlinal cell walls fail to reach the closing stage, and M153, in which development is arrested or strongly retarded after the closing stage. The resulting phenotype of M153 is characterized by a reduction in the overall number of cells, particularly in the aleurone, and unusually thickened cell walls (Olsen *et al.*, 1995).

The homozygous M153 mutant endosperm has been studied by immunohisto-chemistry (Olsen *et al.*, 1995). Abnormalities were observed in the critical first periclinal division of the alveoli which gives rise to the initials from which the starchy endosperm and aleurone develop. Many phragmoplasts were abnormally oriented or appeared fragmented, suggesting disturbances in the control of the division plane as well as in the organization and/or expansive growth of the phragmoplasts. Similar studies are currently in progress with other Type I mutants. Such data will contribute significantly to our understanding of the mechanisms involved in barley wild-type endosperm development.

Mutant phenotypes of Type II barley endosperm display irregular patterns of endosperm development. This group includes B10, which exhibits unorganized callus-like cell populations in addition to the usual starchy endosperm and aleurone, N17 and B13, which have separate endosperm wings and lack the dorsal prismatic starchy endosperm cells, and N34, in which either the left or the right endosperm half fails to develop normally, resulting in an asymmetrical endosperm.

Phenotypes of mutants such as N17 and N34 suggest that the two 'lobes' of irregular starchy endosperm cells are derived from separate lineages of nuclei, established by the two daughter nuclei resulting from the first mitotic division of the primary endosperm nucleus (see Figure 9 in Bosnes *et al.*, 1992). A similar highly regulated control of division plane and migration 'routes' for the initial endosperm nuclei in maize endosperm has been postulated to explain the patterns of revertant colour sectors in unstable genotypes (reviewed by Walbot, 1994).

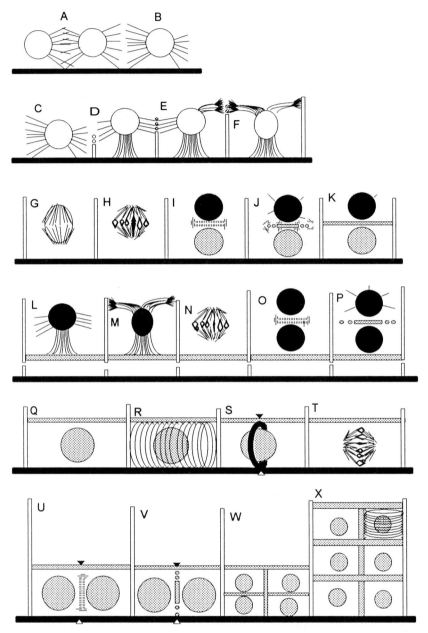

Figure 10.2. *Summary of key events in the development of barley endosperm. Syncytial stage (0–5 DAP) A–B. Mitosis without cytokinesis populates the endosperm and nuclear-based radial microtubules organize the cytoplasm into nuclear-cytoplasmic domains (NCDs). Cellularization (5–8 DAP) C–K. In the first phase (C–F), mitosis is halted as NCDs are polarized in association with specialized microtubule systems; anticlinal walls are deposited at boundaries of polarized NCDs. Cytoplasmic phragmoplasts form at interfaces of opposing*

Future studies on barley will focus on identification of the cellular mechanisms underlying these important developmental phenomena.

In the mature wild-type endosperm, the starchy endosperm consists of three main cell types, namely the irregularly shaped cells that become filled with starch in the interior of the grain, the prismatic cells that appear to fan out from a region located over the ventral crease area, and the subaleurone cells that develop adjacent to the aleurone layer. Mutants in which prismatic cells are entirely lacking, suggest that the prismatic cells are derived from a zone of mitotic activity over the crease area (Bosnes *et al.*, 1992). Such localized mitotic activity has yet to be confirmed by direct observation. As is outlined in Figure 10.2, the cells destined to become aleurone cells are set aside immediately after the formation of the first periclinal cell wall. A few days later the subaleurone cells develop at the interface between aleurone and starchy endosperm. It is our view that the subaleurone cells are derived from the aleurone cell meristem by periclinal divisions, and that they are subsequently redifferentiated to the starchy endosperm cell phenotype. Redifferentiation involves loss of aleurone grain vacuoles and development of the capacity to synthesize hordein (Bosnes *et al.*, 1992). Although no direct evidence exists for this process in barley, studies on wheat have shown that periclinal divisions in the peripheral aleurone cells give rise to the inner cell type which subsequently loses both the characteristic autofluorescent wall components and the small Type I aleurone cell vacuolar inclusions (Morrison *et al.*, 1975). The ontogeny of the subaleurone cells in barley will be re-examined using newly developed immunohistochemical methods for the study of developing endosperm (Brown *et al.*, 1994).

10.6 Developmental genetics

It is assumed that a hierarchy of endosperm-specific genes are sequentially transcribed and that their expression serves to regulate the different developmental phases (Olsen *et al.*, 1992). Examples of key regulatory genes may include those involved in the suppression of functional interzonal phragmoplasts and/or cell

---

microtubule systems (F). In the second phase (G–K), mitotic activity resumes, followed by typical cytokinesis resulting in the first periclinal walls. The daughter cells from this division have different developmental fates; peripheral cells (stippled) give rise to initials of the aleurone layer, while those in the interior (solid) develop into the starchy endosperm. Differentiation (8–21 DAP) L–X. Two co-ordinated processes lead to formation of the starchy endosperm. Compartmentalization into alveoli (L–P) proceeds centripetally until the entire central endosperm is cellularized at about 8 DAP. Subdivision into smaller cells by mitosis followed by cytokinesis continues from 8 to 14 DAP (L–P). Aleurone cell differentiation (Q–X) is initiated more or less synchronously in the peripheral layer of cells. The three-layered peripheral aleurone develops from 8 to 21 DAP. Interestingly, the microtubule cycle in this tissue is typical of plant meristems with hoop-like cortical microtubule arrays in interphase (R) and PPBs predicting future division planes (S).

plates at the syncytial stage, the initiation of the free-growing cell wall, the activation of functional interzonal phragmoplasts and cell plates, and the activation of PPBs in aleurone cell differentiation (see Figure 10.2). It is likely that the key regulatory genes can be identified through their mutant phenotypes. Several of the Type I barley mutants described above are possible candidates. In future work, a priority will be the design of criteria for the isolation of relevant mutant phenotypes which can serve as the basis for isolation of major developmental genes either by transposon tagging or by marker-assisted cloning strategies. Assuming that such genes are expressed in an endosperm-specific manner, one strategy for cloning the genes would be the identification of mutants in which only the endosperm (and not the embryo) is affected. Although mutants in which the main effect is on the endosperm do exist in maize, the majority of *dek* mutants in maize (Sheridan *et al.*, 1986) and *sex* mutants in barley (Bosnes *et al.*, 1987) affect both embryo and endosperm, and are most likely to be mutants of housekeeping genes.

When designing strategies for mutant screens, one should consider the possibility that a single gene may play a central role in the development of both endosperm and embryo. Two lines of evidence suggest the existence of genes with overlapping patterns of expression. The first example is the maize *dek*1 mutant. In *dek*1 the endosperm lacks aleurone cells altogether, and the embryo, which is highly disorganized, lacks a shoot meristem (Dolfini and Sparvoli, 1988; Sheridan *et al.*, 1986). Although the *dek*1 gene has not been cloned, the distinctive effects of the mutant gene in embryo and endosperm indicate that a direct genetic effect is a likely explanation. The second line of evidence for genes with an overlapping pattern of expression in the embryo and endosperm derives from the isolation of cDNA clones expressed in developing aleurone layers and in embryos (Aalen *et al.*, 1994). Definite proof that these transcripts are encoded by the same genes must await analysis of the corresponding gene promoters in transgenic cereal plants.

10.7 Identification of transcripts expressed during barley endosperm development by differential screening

10.7.1 *Transcripts expressed at the syncytial and cellularization stages*

The coenocytic endosperm, which develops deep within the protective layers of the ovule and ovary, is not readily accessible for direct biochemical or molecular studies. Using *in situ* hybridization, Dow and Mascarenhas (1991a, b) observed a high rate of accumulation of ribosomes in the central cell of mature embryo sacs. In addition, Bosnes and Olsen (1992) presented evidence that the rate of gene transcription increases sixfold during the syncytial stage. Taken together, these observations indicate that the endosperm coenocyte is a rich source of poly(A)rich RNA from which it should be feasible to isolate endosperm-specific cDNAs shortly after fertilization. We therefore designed a differential screening strategy in which a lambda ZAPII cDNA library constructed from intact ovaries at 5 DAP (Figure 10.3a) was screened with a positive probe based on manually dissected embryo sacs with adhering nucellus and integuments (Figure 10.3b) and a negative probe from isolated pericarp (Doan, Linnestad and Olsen, unpublished data). The pericarp is a relatively complex tissue consisting of an outer epidermis, several layers of starch-producing parenchyma cells, and two layers of cross cells, the only chlorenchymatous cells of the ovary. At 5 DAP, the endosperm is in the syncytial stage as shown in Figure 10.2a. In addition to the endosperm coenocyte, the positive probe contained poly(A)rich RNA from the degrading nucellus as well as from the integuments. The positive probe was therefore expected to recognize not only clones representing endosperm transcripts, but also plaques containing cDNAs representing transcripts from the two sporophytic tissues.

In the differential screening experiment, the positive and the negative probes gave a strong or intermediate hybridization signal with 20% and 40% of the plated plaques, respectively. In total, 1.5% of clones in the cDNA library represented transcripts expressed in the endosperm coenocyte, the nucellus or the integuments, but not in the pericarp. After isolation of inserts, rescreening with both probes and Southern hybridization, the positive clones selected for further analysis fell into 32 homology classes. Of these groups, one consisted of five individual clones, one group contained four clones, two had three clones, and two groups consisted of two clones each. The rest of the positive cDNAs were individual clones which did not cross-hybridize with any of the other clones. In order to investigate the cellular specificity of the transcripts within the tissues of the positive probe, a preliminary *in situ* hybridization analysis was carried out on transverse sections of 5 DAP ovules with four randomly selected clones. The results obtained indicate that two of these clones represents a transcript detectable only in the endosperm coenocyte, whereas the other two are expressed in the nucellus and the nucellar projection. To our knowledge, this is the first report

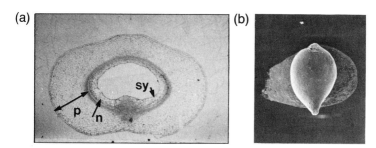

Figure 10.3. *Transverse section of 5 DAP barley ovary consisting of (a) co-enocytic endosperm (sy), nucleus and testa cell layers (n) and the pericarp (p); (b) band dissected 5 DAP embryo sac with adhering nucleus and testa cell layers.*

of the isolation of a cDNA clone for a transcript that is differentially expressed during the initial stages of cereal endosperm development. The presence of an endosperm-specific cDNA clone supports the conclusion of Bosnes and Olsen (1992) that the endosperm coenocyte is actively engaged in RNA transcription. Thus, the presence of poly(A)rich RNA in the nucellus and testa of the positive probe does not cause a problem in the isolation of endosperm-specific cDNA. On the contrary, the presence of transcripts from these sources facilitates the isolation of cDNA clones from the sporophytic tissues immediately surrounding the embryo sac, tissues that are important in the development of the megagameto-phyte as well as the endosperm. As indicated by [^{3}H]-uridine uptake by young barley ovules, the nucellus and nucellar projection are actively engaged in RNA synthesis, and it is therefore not surprising that such clones should appear among the positive clones (Bosnes and Olsen, 1992).

In the future, several strategies will be employed to identify the function of the numerous endosperm-specific transcripts expected to be found in our cDNA collection. First, antibodies directed against the proteins encoded by the clones will be used as probes for subcellular localization, thereby suggesting protein function. Secondly, the pattern of expression of the transcripts will be examined in our collection of barley endosperm *sex* mutants. Reverse genetics experiments similar to those conducted on the maize *ent-kaurene* gene by Bensen *et al.* (1995), in which mutator-induced mutants were identified by a PCR-based strategy, will be used to study the function of the transcripts. In addition, the activity of the promoters of the corresponding genes will be studied in transgenic rice in order to identify the regulatory element responsible for early endosperm gene expression (see Kalla *et al.*, 1994).

10.7.2 *Aleurone cell-specific transcripts*

Aleurone cell differentiation is a process that is strictly regulated in time and space. The peripheral layer of aleurone is initiated by the first periclinal divisions in the alveoli, and the aleurone begins to differentiate at around 8 DAP (Figure 10.2). In order to identify the transcripts and gene promoters involved in the specification of aleurone cell differentiation, a cDNA library constructed from

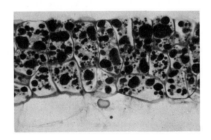

Figure 10.4 *Portion of isolated barley aleurone layer 20 DAP.*

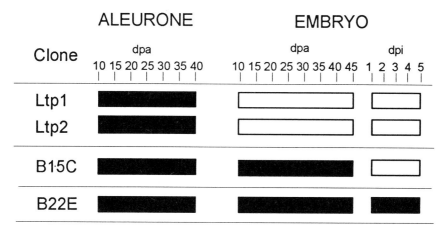

Figure 10.5. *Expression patterns of aleurone-positive cDNA clones.*

poly(A)rich RNA of endosperm isolated at 20 DAP was screened with a positive probe derived from aleurone layers isolated at 20 DAP (Figure 10.4), and a negative probe derived from extruded starchy endosperm at the same stage (Jakobsen *et al.*, 1989). On the basis of cross-hybridization data, as well as Northern and *in situ* hybridization analyses, the positive clones resulting from this experiment have been shown to fall into three main subgroups (Aalen *et al.*, 1994) (Figure 10.5). Of these, only the first group, including the non-specific lipid transfer proteins Ltp1 and Ltp2, represent authentically aleurone-specific transcripts. In the second group, represented by B15C, which has recently been identified as a peroxydoxin (R. Aalen, personal communication), expression is seen in the embryo as well as in the endosperm. For the clones of the last subgroups, represented by the B22E clone, transcripts are present in germinating embryos as well (Klemsdal *et al.*, 1991; Olsen *et al.*, 1990). The first of the cloned transcripts to appear in the developing grain is *Ltp2*, which is expressed shortly after the onset of aleurone cell differentiation, and is detectable at 9 DAP by *in situ* hybridization (Kalla *et al.*, 1994). The promoter of this gene was chosen for an analysis of promoter elements involved in early aleurone cell transcription. A construct containing the *Ltp2* promoter fused to the *Gus* gene has been transformed to rice (Kalla *et al.*, 1994). In these rice plants, *Gus* activity is detectable exclusively in the aleurone cells 1 day after the onset of aleurone cell differentiation, corresponding closely to the temporal regulation of *Ltp2* transcription in barley. In rice, the initial *Gus* activity is located dorsally, spreads laterally, and can eventually be detected throughout the aleurone layer. In barley, the lack of a routine transformation protocol has prevented the undertaking of similar studies, but it is tempting to speculate that a similar dorso-ventral gradient exists in barley. Interestingly, this gradient is the reverse of the gradient observed in aleurone cell differentiation, which starts in the ventral crease area and spreads to the dorsal surface. Sequence alignment has identified putative binding sites for *myb* and *myc* transcription factors in an arrangement similar to that of the anthocyanin structural gene *Bz1-McC* of maize (Roth *et al.*, 1991). The *Bz1-McC* gene is thought to be activated in aleurone cells by the binding of the *myb* and *myc* homologues *C1* and *Lc*. The possibility that similar transcription factors are involved in the activation of *Ltp2* transcription is currently being investigated. Probes representing transcripts expressed in the early phase of aleurone cell differentiation will be used in the characterization of barley *sex* mutants. In particular, the *Ltp2* probe should yield interesting results in mutant B9, which lacks aleurone cells.

10.8 Progress and perspectives

This chapter has been written from the perspective that the study of endosperm development can contribute to the overall understanding of embryogenesis. Just as examination of the phenotype of mutant genes can provide insights into the events controlled by such genes, so investigation of the unusual development of endosperm can provide clues as to the fundamental controls of embryo development. Similarly, an understanding of the genetic reprogramming which may have led to the evolution of the endosperm may aid identification of central genes in embryogenesis. Regardless of the evolutionary origin of the endosperm, many of the basic processes (i.e. establishment of polarity and determination of division plane, function of cytoskeletal components and the complex mechanisms of cell wall formation) which are essential for embryogenesis can be more easily studied in the endosperm than in the embryo itself.

Plant embryologists have isolated many mutants with impaired developmental patterns with the aim of isolating the genes that control embryo development. As more information about the development of the embryo and endosperm becomes available, it will be valuable to compare the effect of mutant genes in the two systems in order to evaluate the specificity of the effect of the mutant genes. In both systems even housekeeping genes can cause interesting mutant phenotypes, and there are known cases of a mutant gene causing different phenotypes in embryo and endosperm. For example, the *dek*1 gene in maize causes failure of aleurone development in the endosperm and failure of a shoot meristem in the embryo. Possibly the greatest challenge which lies ahead concerns the deciphering of the respective roles of such genes with dual phenotypic expression. The close-knit developmental relationships of the embryo and endosperm provide an unprecedented opportunity to analyse the genetic control of the essential processes of plant development.

References

Aalen, R.B., Opsahl-Ferstad, H.G., Linnestad, C. and Olsen, O.-A. (1994) Transcripts encoding an oleosin and a dormancy-related protein are present in both the aleurone layer and the embryo of developing barley (*Hordeum vulgare* L.) seeds. *Plant J.* **5**, 385–396.

Battaglia, E. (1980) Embryological questions. 2. Is the endosperm of angiosperms sporophytic or gametophytic? *Ann. Bot. (Roma)* **39**, 9–30.

Bensen, R.J., Johal, G.S., Crane, C.C., Tossberg, J.T., Schnabel, R.B.M. and Briggs, S. (1995) Cloning and characterization of the maize *An1* gene. *Plant Cell* 7, 75–84.

Bosnes, M. and Olsen, O.-A. (1992) The rate of nuclear gene transcription in barley endosperm syncytia increases sixfold before cell-wall formation. *Planta* 186, 376–383.

Bosnes, M., Harris, E., Ailgeltinger, L. and Olsen, O.-A. (1987) Morphology and ultrastructure of 11 barley shrunken endosperm mutants. *Theor. Appl. Genet.* 74, 177–187.

Bosnes, M., Weideman, F. and Olsen, O.-A. (1992) Endosperm differentiation in barley wild-type and *sex* mutants. *Plant J.* 2, 661–674.

Briggs, S. (1995) Cloning and characterization of the maize *An1* gene. *Plant Cell* 77, 75–84.

Brown, R.C., Lemmon, B.E. and Olsen, O.-A. (1994) Endosperm development in barley: microtubule involvement in the morphogenetic pathway. *Plant Cell* 6, 1241–1252.

Clark, J.K. and Sheridan, W.F. (1986) Developmental profiles of the maize embryo-lethal mutants *dek 22* and *dek 23*. *J. Hered.* 77, 83–92.

Dolfini, S.F. and Sparvoli, F. (1988) Cytological characterization of the embryo-lethal mutant *dek-1* of maize. *Protoplasma* 144, 142–148.

Dow. D.A. and Mascarenhas, J.P. (1991a) Optimization of conditions for *in situ* hybridization and determination of the relative number of ribosomes in the cells of the mature embryo sac of maize. *Sex Plant Reprod.* 4, 244–249.

Dow, D.A. and Mascarenhas, J.P. (1991b) Synthesis and accumulation of ribosomes in individual cells of the female gametophyte of maize during its development. *Sex Plant Reprod.* 4, 250–253.

Favre-Ducharte, M. (1984) Homologies and phylogeny. In: *Embryology of Angiosperms* (ed. B.M. Johri). Springer-Verlag, Berlin, pp. 697–734.

Felker, F.C., Peterson, D.M. and Nelson, O.N. (1985) Anatomy of immature grains of eight maternal effect shrunken endosperm barley mutants. *Am. J. Bot.* 72, 248–256.

Friedman, W.E. (1992) Evidence of a pre-angiosperm origin of endosperm: implications for the evolution of flowering plants. *Science* 255, 336–339.

Friedman, W.E. (1994) The evolution of embryology in seed plants and the developmental origin and early history of endosperm. *Am. J. Bot.* 81, 1468–1486.

Gifford, E.M. and Foster, A.S. (1989) *Morphology and Evolution of Vascular Plants*, 3rd edition. W.H. Freeman and Company, New York.

Huang, B.Q. and Russell, S.D. (1992) Female germ unit: organization, isolation, and function. *Int. Rev. Cytol.* 140, 233–293.

Jakobsen, K., Klemsdal, S., Aalen, R., Bosnes, M., Alexander, D. and Olsen, O.-A. (1989) Barley aleurone cell development: molecular cloning of aleurone-specific cDNAs from immature grains. *Plant Mol. Biol.* 12, 285–293.

Kalla, R., Shimamoto, K., Potter, R., Nielsen, P.S., Linnestad, C. and Olsen, O.-A. (1994) The promotor of the barley aleurone cell specific expression in transgenic rice. *Plant J.* 6, 849–860.

Klemsdal, S.S., Hughes, W., Lönneborg, A., Aalen, R.B. and Olsen, O.-A. (1991) Primary structure of a novel barley gene differentially expressed in immature aleurone layers. *Mol. Gen. Genet.* 228, 9–16.

Lowe, J. and Nelson, O.E. (1946) Miniature seed — a study in the development of a defective caryopsis in maize. *Genetics* 31, 525–533.

Mangelsdorf, P.C. (1926) The genetics and morphology of some endosperm characters in maize. *Conn. Agric. Exp. Stn. Bull.* 279, 513–612.

Morrison, I.N., Kuo, J. and O'Brien, T.P. (1975) Histochemistry and fine structure of developing wheat aleurone cells. *Planta* 123, 105–116.

Neuffer, M.G. and Sheridan, W.F. (1980) Defective kernel mutants of maize. I. Genetic and lethality studies. *Genetics* **95**, 929–944.

Olsen, O.-A., Jakobsen, K.S. and Schmelzer, E. (1990) Development of barley aleurone cells: temporal and spatial patterns of accumulation of cell-specific mRNAs. *Planta* **181**, 462–466.

Olsen, O.-A., Potter, R.H. and Kalla, R. (1992) Histo-differentiation and molecular biology of developing cereal endosperm. *Seed Sci. Res.* **2**, 117–131.

Olsen, O.-A., Brown, R.C. and Lemmon, B.E. (1995) Pattern and process of wall formation in developing endosperm. *BioEssayss* **17**, 803–812.

Ramage, R.T. and Crandall, C.L. (1981) Defective endosperm xenia (*dex*) mutants. *Barley Genet. Newslett.* **11**, 32–33.

Roth, B.A., Goff, S.A., Klein, T.M. and Fromm, M.E. (1991) *C1*- and *R*-dependent expression of the maize *Bz1* gene requires sequences with homology to mammalian *myb* and *myc* binding sites. *Plant Cell* **3**, 317–325.

Sheridan, W.F. and Neuffer, M.G. (1980) Defective kernel mutants of maize. II. Morphological and embryo culture studies. *Genet.* **95**, 945–960.

Sheridan, W.F. (1988) Maize developmental genetics: genes of morphogenesis. *Annu. Rev. Genet.* **22**, 353–385.

Sheridan, W.F., Clark, J.K., Chang, M.T. and Neuffer, M.G. (1986) The *dek* mutants — new mutants defective in kernel development. *Maize Genet. Coop. Newslett.* **60**, 64.

Walbot, V. (1994) Overview of key steps in aleurone development. In: *The Maize Handbook* (eds M. Freeling and V. Walbot). Springer-Verlag, New York, pp. 78–80.

Yakovlev, M.S. (1967) Polyembryony in higher plants and principles of its classification. *Phytomorphology* **17**, 278–282.

Initiation of somatic embryos from single cells

M.A.J. Toonen and S.C. de Vries

11.1 Introduction

Somatic or asexual embryogenesis is the process whereby somatic cells develop into plants via characteristic morphological stages. The later stages in particular closely resemble zygotic embryo development, as evidenced by the familiar globular, heart-shaped and torpedo-shaped stages in dicotyledons. In mono-cotyledons, the globular stage develops into a transition-state embryo which contains an 'integral' suspensor. A scutellum is formed at a lateral position of the transition-state embryo, and shoot and root primordia develop at both ends of the embryo axis. The scutellum subsequently develops into the single cotyledon. Yet another type of embryo development takes place in conifers (Tautorus *et al.*, 1991). Furthermore, in monocotyledons and conifers the morphology of developing somatic embryos from the globular stage onwards closely resembles that of their zygotic counterparts.

In this chapter we will focus on the first phase of somatic embryogenesis, the transition of somatic cells into cells, referred to as embryogenic cells, that will eventually develop into the globular embryo (Figure 11.1). The term 'embryogenic cell' will be restricted to those cells that have completed the transition from a somatic state to one in which no further externally applied stimuli are necessary to produce the somatic embryo (De Jong *et al.*, 1993b). The cells that are in this transitional state and have started to become embryogenic, but still require externally applied stimuli, are defined as competent cells (Toonen *et al.*, 1994). Explant cells can be triggered to become embryogenic by a variety of procedures that usually include exposure to plant growth regulators; pH shock, heat shock and treatment with various chemicals have also been reported. In general, only a very limited number of cells in any given explant respond by becoming embryogenic. Between the time when the inducing stimulus is applied and the appearance of the first morphologically visible sign of the formation of a somatic embryo, there is emerging experimental evidence which suggests that a number

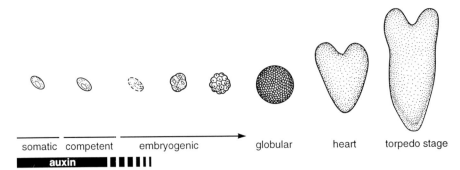

Figure 11.1. *Schematic representation of indirect somatic embryogenesis. The somatic cell stage before induction, the competent cell stage in the presence of auxin and the embryogenic cell stage in the absence of externally applied growth regulators are shown. In direct somatic embryogenesis, somatic cells develop directly into somatic embryos, probably without an intervening, embryogenic cell state.*

of different processes take place. In order to determine the order of these different processes, it is important to describe first a number of experimental systems that allow the study of the direct formation of somatic embryos from single explant cells. The alternative, namely the formation of somatic embryos from single cells produced via an intermediate embryogenic callus phase, is more complex to interpret in terms of individual cell behaviour. However, it may well be that the processes involved are similar.

11.2 Single-cell origin of somatic embryos

There are a number of reasons why it is desirable to have an experimental system that allows the direct development of somatic cells into somatic embryos, or at least allows a single cell-specific recording of the process. Firstly, and most importantly, the changes that are observed in a particular cell can only be correlated directly with the process of embryogenesis if there is irrevocable proof that it is indeed the particular cell in which the change is observed that develops into an embryo. Secondly, somatic embryos that are derived from single cells usually resemble the zygotic ones more closely in morphology and quality, a trait that is of paramount importance to applications of somatic embryos in plant propagation. Thirdly, descriptions of cellular events and the interpretation of molecular markers are greatly simplified when dealing with somatic embryogenesis starting from a single cell.

11.2.1 *Direct somatic embryogenesis*

In *Cichorium* spp., most somatic embryos arise from fully differentiated mesophyll cells and vascular sheath parenchyma cells. Five days after induction, modified cells are visible that show a reduction in the amount of cytoplasm, loss of plastids, an enlarged nucleus surrounded by a large vacuole and thick radial strands of cytoplasm. These changes are preceded by the deposition of callose in the cell wall of the modified cells (Figure 11.2). Furthermore, newly formed cell walls show callose deposition (Dubois *et al.*, 1991). No polar development of the embryo is observed before the late globular stage, and a suspensor is not formed during the early divisions. In *Trifolium repens*, somatic embryos also develop from single cells or small clusters of cells which are surrounded by a callose-rich cell wall (Maheswaran and Williams, 1985).

Mesophyll cells of leaf segments of the monocotyledon *Dactylis glomerata* divide after 4 days of induction. In 90% of cases this division is periclinal to the leaf surface. For the following divisions, three different division patterns have been observed. In one case the apical cell divides to form the embryo, while the basal cell develops into a suspensor-like structure. The other types of division lead to a cell cluster that develops into a somatic embryo or an embryogenic cell mass (Trigano *et al.*, 1989). In *Cichorium* as well as in *Dactylis* a minority of the epidermal cells are also capable of somatic embryo development. In *Medicago sativa* somatic embryos are derived only from epidermal cells of the hypocotyl. The cells are identified as isodiametric or palisade-shaped cells that undergo a periclinal division leading to the formation of an apical and a basal cell. The second division is again periclinal and occurs in the basal or apical cell. This three-celled

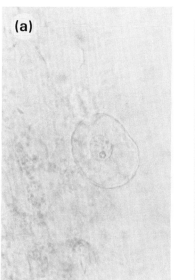
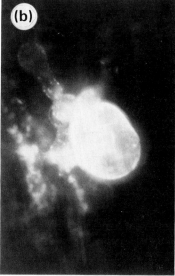

Figure 11.2. *Direct somatic embryogenesis starting from a single epidermal cell of a* Cichorium *leaf explant. (a) Bright field image of an activated leaf explant. (b) Fluorescent image of the same cell stained with aniline blue indicating the presence of a callose-containing wall. (Photographs kindly provided by Dr Jean-Louis Hilbert, University of Lille, France.)*

structure is also observed in zygotic embryos. A similar development of single cells has been observed in cotyledons and roots of induced plantlets of *Medicago sativa* (Dos Santos *et al.*, 1983). Furthermore, somatic embryos from *Ranuculus sceleratus* may develop from epidermal cells. In stems, small cytoplasm-rich cells are present that are smaller than the other epidermal cells. These cells contain plastids which have starch grains and scattered ribosomes. A symmetrical division occurs, leading to the formation of an embryogenic two-celled cluster (Konar *et al.*, 1972).

The development of somatic embryos directly on the surface of zygotic embryos is frequently observed. The basis of this observation is not known, but the cells of the zygotic embryo are thought to be 'young' and developmentally related to the totipotent zygote. Somatic embryos from *Oryza sativa* develop from single scutellum epithelial cells. These cells undergo a periclinal division, resulting in the formation of a basal and a terminal cell. The terminal cell divides longitudinally, but transverse or oblique divisions are also observed (Jones and Rost, 1989a). Single cells from cryopreserved somatic embryos of *Picea sitchensis* can easily develop into somatic embryos. Isodiametric cells with an enlarged nucleus divide 24 h after thawing. The first divisions are symmetrical, leading to the formation of a cell cluster of four to eight cells. After 5 days this cluster dissociates from the somatic embryo and polar growth is established by vacuolation and the outgrowth of suspensor-like cells (Kristensen *et al.*, 1994).

11.2.2 *Indirect somatic embryogenesis*

Since the first report of Reinert, in 1959, that somatic embryos of *Daucus carota* may have a single cell origin, a number of reports on this subject have been published. Free-floating cells released from callus (Kato and Takeuchi, 1963) and single cells isolated from suspension cultures are able to develop into somatic embryos (Backs-Hüsemann and Reinert, 1970; Nomura and Komamine, 1985). Suspension cultures initiated from hypocotyls are widely used as a model system for somatic embryogenesis (Zimmerman, 1993). After 5 to 7 days of auxin treatment, epidermal and cortical cells of the hypocotyl detach from the central cylinder, and in contrast to the provascular cells, these cells do not proliferate. Provascular cells show an increase in the amount of cytoplasm, and the vacuolar volume decreases. After longitudinal or transverse divisions, small isodiametric cells are formed (Guzzo *et al.*, 1994). These cells elongate, and after removal of auxin only a particular oval intermediate state of these cells develops into somatic embryos as observed by cell tracking (F. Guzzo, M. Toonen and S. C. de Vries, unpublished data). None of the other cell types, including the small isodiametric rapidly dividing cells in the so-called starting cell culture that was formed this way, are seen to develop into somatic embryos. However, in an established suspension culture, somatic embryos develop from morphologically quite different single cells, including small cytoplasmic cells, and elongated vacuolated cells passed through a 22-μm mesh (Toonen *et al.*, 1994 and Figure 11.3). These results were obtained by cell tracking of over 30 000 individual cells, and they also reveal

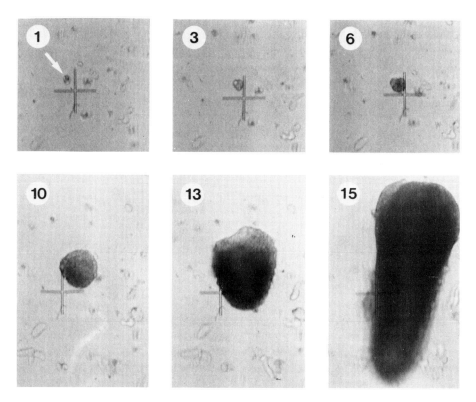

Figure 11.3. *Indirect somatic embryogenesis in* Daucus carota *viewed by semi-automatic cell tracking. Only one single cell present at day 1 (arrow) developed via an embryogenic cluster (days 3 and 6) into a globular (day 10), heart-shaped (day 13) and torpedo-shaped (day 15) embryo. For experimental details see Toonen* et al. *(1994).*

great variability in the early cell divisions that lead to the formation of somatic embryos.

The single cell origin of somatic embryos in suspension cultures has been reported in a number of other plants. However, no direct proof has been obtained by cell-tracking or by time-lapse photography, and the evidence for these reports was based mainly on the observation that single cells divide asymmetrically, like the zygote (Karlsson and Vasil, 1986b; Konar *et al.*, 1972; Vieitez *et al.*, 1992).

Indirect somatic embryo development in monocotyledons shows a general pattern. Cells that develop into somatic embryos are located at the periphery of the callus, they are rich in cytoplasm, form a thick cell wall and undergo a series of internal cleavage-type divisions. This pattern has been reported to occur in *Panicum maximum* (Lu and Vasil, 1985) and *Saccharum officinarum* L. (Ho and Vasil, 1983). A suspensor-like structure has been observed in somatic embryos of *Avena sativa*, where an asymmetrical division is followed by cleavage-type divisions in one of the daughter cells (Chen *et al.*, 1994). Many somatic embryos of *Pennisetum americanum* also contain a suspensor-like structure which connects them to the callus (Botti and Vasil, 1984; Vasil and Vasil, 1982).

Single cells of *Picea abies* and *P. glauca* hypocotyl callus develop into somatic embryos after unequal division. A distal small cell with dense cytoplasm is formed, which develops into a somatic embryo after periclinal and anticlinal divisions. The basal cell divides to form two to four suspensor cells (Nagmani *et al.*, 1987). Single cells in friable *Medicago sativa* callus show a pattern of development similar to that of *Medicago* epidermal cells (Dos Santos *et al.*, 1983).

From these studies it is difficult to draw definite conclusions concerning the importance of the first divisions in the development of somatic cells into embryogenic cells. There appears to be some evidence for a pattern similar to that observed in zygotic embryogenesis, but this does not appear to be essential, because in many instances somatic embryos form without a clear regularity in the first divisions, and indeed most do not have suspensor-like appendages. The first changes observed in single activated cells often appear to affect the wall composition and internal organization, well before the first division takes place in the activated cells.

11.3 Cellular aspects of somatic embryo initiation

11.3.1 *The inducing treatments*

In most cases the herbicide 2,4-dichlorophenoxyacetic acid (2,4-D) is used to induce the formation of somatic embryos from explant cells, as in *Avena sativa* (Chen *et al.*, 1994), *Daucus carota* (Toonen *et al.*, 1994), *Medicago sativa* (Dos Santos *et al.*, 1983), *Oryza sativa* (Jones and Rost, 1989a), *Pennisetum americanum* (Vasil and Vasil, 1982), *Saccharum officinarum* (Ho and Vasil, 1983), *Sorghum bicolor* (Wernicke and Brettell, 1980) and *Tylophora indica* (Rao and Narayanaswami, 1972). Furthermore, combinations of 2,4-D and cytokinins, such as zeatin in *Daucus carota* (Nomura and Komamine, 1985) and benzyladenine (BAP) in *Fagus sylvatica* (Vieitez *et al.*, 1992) and *Picea glauca* (Attree *et al.*, 1990), have been successively used. Microcallus suspensions of *Medicago sativa* maintained on a less stable synthetic auxin such as naphthalenacetic acid (NAA) develop into somatic embryos after a short pulse with 2,4-D (Dudits *et al.*, 1991). Embryogenic callus of an embryogenic *Medicago sativa* clone shows an increased sensitivity to 2,4-D compared to a non-embryogenic line derived from the same genotype (Börge *et al.*, 1990). Cell suspensions derived from 2,4-D-treated *Daucus carota* hypocotyls can also be maintained in the absence of auxin at a low pH. After an increase in pH, somatic embryos develop from the cell clusters (Smith and Krikorian, 1990). This effect is thought to mimic certain effects of auxins on hyperpolarization of the cell membrane and the cytoplasmic and cell wall pH, which have been reviewed by Dudits *et al.* (1991, 1993). Other hormones, such as dichloro-*o*-anisic acid (Dicamba), have been applied to *Dactylis glomerata* (Trigano *et al.*, 1989). Glycerol is used as an inducer of somatic embryogenesis in *Citrus auranitum* (Gavish *et al.*, 1992). Glycerol in combination with a mixture of the growth regulations NAA and N^6-(2-isopentenyl)adenine is used to stimulate direct embryogenesis from *Cichorium* leaves (Hilbert *et al.*, 1992).

There are also reports which indicate that the frequently used compounds such as 2,4-D are not unique in stimulating the formation of embryogenic cells. Somatic embryo initiation on megagametophyte tissue of *Pinus taeda*, for instance, does not require externally applied growth regulators, a finding that can be explained by assuming that the megagametophyte tissue itself supplies sufficient phyto-hormones for somatic embryo initiation (Becwar *et al.*, 1990). Spontaneous formation of somatic embryos from cryopreserved somatic embryos of *Picea sitchensis* occurs after thawing (Kristensen *et al.*, 1994). Other well-known examples where somatic embryogenesis takes place without any growth regulators as inducing agent are found in microspore systems in several species. In late microspore and early binucleate pollen, the developmental pathway leading to the formation of mature pollen can be redirected towards embryogenesis by heat shock (in *Brassica*) (Duijs *et al.*, 1992; Swanson *et al.*, 1987) or starvation (in *Nicotiana tabacum*) (Zársky *et al.*, 1992).

The occurrence of somatic embryos in plants is widespread, and their induction can be spontaneous at one end of the spectrum and may require long and complex treatments with growth regulators and other compounds at the other end. The range of possible induction treatments suggests that it is unlikely that a single inducing molecule is responsible. In the following sections a number of cellular events that are often found to be associated with the first stages of somatic embryo induction will be discussed in some detail.

11.3.2 *Plane of division*

In general, the first division of the zygote is asymmetrical. The first stages of zygotic embryogenesis have been well studied in *Arabidopsis thaliana* (Mansfield and Briarty, 1991) and *Capsella bursa-pastoris* (Schulz and Jensen, 1968). In these plants, the zygote divides transversely to form a basal vacuolated daughter cell that is larger than the apical cell. The basal cell develops into the suspensor by transverse divisions, and the top cell of the suspensor will develop into the hypophysis. The cytoplasm-rich apical cell divides longitudinally, and this division is followed by a second division perpendicular to the plane of the first division. In *Arabidopsis* this pattern of cell division is virtually invariant. Impaired development of the zygotic embryo in the *gnom* mutant of *A. thaliana* is preceded by an aberrant plane and position of the first division (Mayer *et al.*, 1993). This suggests that a correct asymmetrical first division is important for the subsequent development of the zygotic embryo. However, other plants display much more variation in the first division, and also develop normal zygotic embryos. For instance, in *Oryza sativa* the zygote elongates first after fertilization and then divides unequally and transversely. The first division of the smaller terminal cell is often oblique, but may also be transverse or longitudinal (Jones and Rost, 1989b). A similar developmental pattern has been observed during the early stages of somatic embryogenesis. The initial cell division sequence does not follow any particular pattern, nor does it predict subsequent embryo development (Jones and Rost, 1989a). The first division of the zygote of *Daucus carota* is unequal and transverse, and the first division of the apical cell is also transverse (Bothwick, 1931). The first plane of division in *Daucus carota* suspension cells which are developing into somatic embryos is variable. In a number of cases, asymmetrical divisions have been reported, but other types of division also precede somatic embryo development (Backs-Hüsemann and Reinert, 1970; Guzzo *et al.*, 1994; Toonen *et al.*, 1994). After an asymmetrical division, the development of the somatic embryo resembles that of the zygotic embryo. In the case of a symmetrical division, this pattern more closely resembles the development of the apical cell of the divided zygote (Toonen *et al.*, 1994).

An asymmetrical division in *Medicago sativa* protoplasts leads to somatic embryo development, whereas after a symmetrical division callus is formed (Dijak and Simmons, 1988). However, the first division in somatic embryo development of *Dactylis glomerata* (Trigano *et al.*, 1989) is symmetrical and periclinal, and in *Tylophora indica* (Rao and Narayanaswami, 1972) it is symmetrical and transverse. The plane of the second division is either periclinal or anticlinal in one or both daughter cells of *Dactylis* (Trigano *et al.*, 1989). In *Tylophora* the basal cell undergoes a transverse division, whereas the apical cell divides twice in an oblique fashion. These sequences of divisions are also observed in zygotic embryogenesis of the same species (Rao and Narayanaswami, 1972). Individual cells of *Pennisetum americanum* undergo cleavage divisions to form proembryos of two to eight cells (Botti and Vasil, 1984; Vasil and Vasil, 1982). Division planes in *Cichorium* somatic embryos are variable (Dubois *et al.*, 1991). Upon induction by heat shock of *Brassica napus* haploid microspores, the asymmetrical division changes to a symmetrical division, leading to embryo development (Telmer *et al.*, 1993).

In general, it can be concluded that a correct asymmetrical first division is not an essential requirement for somatic or for zygotic embryogenesis. This does not imply that an asymmetrical distribution of intracellular determinants is not important in the early stages of plant embryogenesis, but rather that such a distribution is not necessarily visibly fixed by an asymmetrical first division. It is of interest that there does appear to be a correlation between the regularity, or lack of it, in the early divisions in zygotic embryogenesis and somatic embryogenesis in the same species (Jones and Rost, 1989a; Rao and Narayanaswami, 1972). It will be interesting to compare somatic embryo development in a species that exhibits strict regularity in zygotic division pattern, such as *Capsella* or *Arabidopsis*.

11.3.3 *Microtubule organization*

Evidence for a possible intracellular redistribution of the cytoskeleton in establishing cell polarity in early embryogenesis has been studied in two systems, with completely different results. Two days after treatment of *Medicago sativa* protoplasts, microtubules were fine, more numerous, and arranged in a disordered network. Untreated protoplasts (which did not develop into somatic embryos) contained thick parallel bundles of microtubules (Dijak and Simmons, 1988). However, in somatic embryogenesis in *Larix*, a random distribution of the microtubules was associated with non-embryogenic growth, while protoplasts derived from embryogenic cultures displayed a parallel orientation of the microtubules (Staxen *et al.*, 1994).

11.3.4 *Cell wall*

The cell wall has an important function in plant development in determining cell size and providing a strong support for the cell. Recent evidence indicates that it may also be involved in determining cell fate by the presence of specific proteins and proteoglycans. Another function of certain highly modified cell walls might be to isolate certain cells from the surrounding cells during the very first events in embryogenic cell formation.

Cell size appears to play an important role in somatic embryogenesis. In direct somatic embryogenesis the embryos are generally formed from small isodiametric cells. Suspension cells that develop into somatic embryos do not expand during the initial stages. In both cases the cell wall must prevent unwanted expansion. A number of enzymes have been identified as candidates for regulation of cell expansion (see Carpita and Gibeaut, 1993; Cosgrove, 1993; Kielisewski and Lamport, 1994). However, the mechanism whereby the expansion of the cell wall is precisely controlled in a co-ordinated manner in the early embryo is unknown. Some clues have been found in *Daucus* somatic embryogenesis. The addition of tunicamycin to a suspension culture leads to inhibition of somatic embryogenesis, possibly as a result of uncontrolled expansion of the outer cells of embryogenic cell clusters. The addition of a 38-kDa peroxidase restores somatic embryogenesis, possibly preventing this expansion (Cordewener *et al.*, 1991).

In *Cichorium* and a number of monocotyledons an increase in cell wall thickness has been observed in embryo-forming cells prior to the first observed division. In *Cichorium* (Dubois *et al.*, 1991) and *Trifolium* (Maheswaran and Williams, 1985) this coincided with the deposition of callose in the cell wall (Figure 11.2). A callose-rich cell wall also surrounds the zygote of *Rhododendron* and *Ledum*, and has been postulated to be involved in determining the size and shape of the zygote. Another possibility is that the callose layer isolates the zygote from maternal tissue, or that it retards the entry of macromolecules (Williams *et al.*, 1984). In somatic embryogenesis, a callose-rich cell wall might form a physical barrier between the competent cell and the surrounding tissue, thought to be essential for establishing the new developmental pathway (Dubois *et al.*, 1991). However, embryo-forming epidermal cells of *Ranuculus scleratus* have been shown to be connected to the adjacent epidermal cells by plasmodesmata (Konar *et al.*, 1972), so whether this is a common property of embryogenic cell formation has yet to be established.

A fascinating result was recently reported for *Fucus* embryo development. When the rhizoid cell of a two-celled embryo of this alga was destroyed by laser microsurgery, but part of its wall was left attached to the thallus cell, the latter expanded to fill the remaining void. Upon prolonged contact with the remaining rhizoid cell wall, the fate of the cell was changed to that of a rhizoid cell. However, when the thallus cell was killed but its wall was left attached to the two-celled embryo, the remaining rhizoid cell expanded and upon contact with the thallus wall acquired the characteristics of a thallus cell. This indicates that stable, non-diffusible cell wall components are able to regulate the cell fate of *Fucus* cells (Berger *et al.*, 1994).

11.3.5 *Cell wall and cell membrane proteins*

Several groups of proteins are present in the cell wall (see Showalter, 1993). The occurrence of a number of glycoproteins has been investigated with a view to their potential use as markers of development. Indeeed, some of the expression patterns of these proteins as determined with the aid of monoclonal antibodies directed against particular epitopes reflect developmental stages underlying the anatomical pattern of the carrot root (Smallwood *et al.*, 1994). The epitopes investigated do not coincide with specific differentiated cell types, but appear to correlate more closely with cell position. Using monoclonal antibodies against these epitopes it was proposed that the fate of a cell which reacts with a particular monoclonal antibody can be predicted before it is determined by morphological criteria (Knox *et al.*, 1989). Several cell wall proteins are also developmentally regulated during sexual reproduction. Expression of the plasma membrane JIM8 epitope in *Brassica* flowers has been found in several cell types in anthers and ovules. Both the egg cell and the sperm cell contain a JIM8 reactive epitope. After fertilization, the two-celled embryo proper and the suspensor of the eight-celled embryo are also reactive with JIM8. In globular-stage embryos, the embryo proper has lost the JIM8 epitope even though it is still present in the suspensor cells (Pennell *et al.*, 1991). A similar model has been postulated for somatic embryo-genesis in *Daucus carota*. Expression of the JIM8 epitope was predicted to coincide with an intermediate stage in development from a single somatic cell to a single embryogenic cell (Pennell *et al.*, 1992). The nature of the relationship between the epitope observed in *Brassica* flowers and the epitope seen in suspension cultures is not yet clear. However, recent experiments employing cell-tracking of JIM8-labelled cells have shown that there is no direct correlation between the presence of the JIM8 epitope on the cell wall and the ability of the cell to develop a somatic embryo (M. Toonen *et al.*, manuscript in preparation). It might still be possible, as suggested previously, that the JIM8 epitope plays a more indirect role in somatic embryogenesis, possibly linked to selective cell death in some of the JIM8-labelled cells (Pennell *et al.*, 1992).

A number of experiments have been described where a more direct role of the cell wall or cell wall-derived proteins in the formation of embryogenic cells has been observed. The addition of conditioned medium to newly initiated suspension cultures of *Daucus carota* stimulates the formation of embryogenic cells (de Vries *et al.*, 1988). However the addition of extracellular glycoproteins to *Citrus* embryo cultures results in suppression of embryo development. In more recent studies, arabinogalactan proteins (AGPs) have been isolated from *Daucus carota* and *Lycopersicon esculentum* seeds, and have been shown to contain different fractions that either promote or suppress embryo development. Using monoclonal antibodies ZUM18 and ZUM15 raised against these fractions, different AGP preparations can be isolated from total carrot seed AGPs. The addition of ZUM18-selected AGPs from either *Daucus* and *Lycopersicon* to an embryo culture increases embryo development two-fold. Addition of ZUM15-selected AGPs leads to a decrease in the number of embryos formed (Kreuger and Van Holst, 1995). The addition of carrot seed AGPs to a non-embryogenic cell line re-induces the embryogenic potential in this cell line (Kreuger and Van Holst, 1993). In *Daucus carota* single cell cultures, the percentage of cells that develop into somatic embryos increases with cell density (Toonen *et al.*, 1994), providing additional evidence for the importance of interactions between cells that may or may not involve cell wall proteins in the determination of the subsequent development of the cell. AGPs present in the cell membrane of *Rosa* spp. might also be directly involved in cell division. The addition of Yariv β-D-glycosyl reagent to suspension cultures of *Rosa* inhibits growth of the culture, probably due to an inhibitory effect on cell division (Serpe and Nothnagel, 1994).

In individual cytoplasm-rich *Brassica napus* protoplasts it has been shown that the frequency of microcallus formation can only be stimulated by co-culturing with a specific vacuolated protoplast type (Spangenberg *et al.*, 1985). These results hint at a highly complex series of interactions between cells in culture, where both division and differentiation of cells may be affected by the culture composition in terms of cell types present. For instance, it is possible that particular non-embryogenic cell types produce compounds that stimulate other cells to become capable of entering the pathway which leads to somatic embryogenesis.

11.3.6 *The role of ploidy and the cell cycle*

On the basis of the observations that polyploid cells do not develop somatic embryos (Coutos-Thevenot *et al.*, 1990), and that certain carrot somatic embryos have been found to be haploid, it has been proposed that chromosome reduction mechanisms play a role in the formation of embryogenic cells (Nuti Ronchi *et al.*, 1992a,b). However, torpedo-stage embryos of *Vitis vinifera* (Faure and Nougarède, 1993) and young somatic embryos of *Abies alba* (Schüller *et al.*, 1989) are all diploid. It is therefore not clear whether the ploidy level of the explant tissue is an essential determinant of the ability to respond to treatments that induce the formation of embryogenic cells.

In *Medicago sativa* microcallus suspensions, somatic embryo development can be initiated with a 1-h 2,4-D treatment. Only 1 day after treatment, histone 3-1 mRNA is present, indicating that many cells have entered the S phase of the cell cycle as a result of this treatment (Kapros *et al.*, 1992). The addition of a specific *Rhizobium* lipo-oligosaccharide (Nod factor), which induces nodule formation in *Medicago sativa*, to microcalli of the same species triggers mitosis in a limited number of cells (Savoure *et al.*, 1994), but no data have been reported for a possible effect on somatic embryo development. The temperature-sensitive mutant (*ts*11) of *Daucus carota* arrests at the globular stage at the non-permissive temperature. The addition of Nod factors restores wild-type development whereas the addition of auxin and 6-BAP does not (De Jong *et al.*, 1993a). Whether the effects of the Nod factors is to promote mitosis in certain ts11 cells is not known.

In leaves of *Zea mays* (Dolezelová *et al.*, 1992) and *Pennisetum purpureum* (Karlsson and Vasil, 1986a; Taylor and Vasil, 1987), no correlation was found between a particular point in the cell cycle and the embryogenic competence of these cells. Pollen-derived embryos of *Nicotiana tabacum* (Zársky *et al.*, 1992) and *Brassica napus* (Binarova *et al.*, 1993) originate from haploid vegetative cells. These cells are normally arrested in the G_1 phase and are terminally differentiated (G_2). Upon induction, these cells can enter the cell cycle again and show transition from the G_1 to the S phase. In *Nicotiana*, this activation cannot be blocked by hydroxyurea, suggesting that the activated vegetative cell enters the cell cycle ahead of the G_1/S control point (Zársky *et al.*, 1992). Microspores of *Nicotiana* in the G_1, S or G_2 phase of the cell cycle all give rise to somatic embryos (Binarova *et al.*, 1993) suggesting that the particular point in the cell cycle at which the inducing stimulus is perceived is not of critical importance in the formation of embryogenic cells. However, in none of the above-mentioned cases has this been demonstrated for individual cells.

11.4 Concluding remarks

At present there is insufficient information available about the molecular details of the pathway leading from somatic cells via competent cells to embryogenic cells for us to form a clear picture of the events that must take place. There is some evidence that changes in cell wall composition and in partitioning of cytoplasmic determinants in explant cells occur as a result of treatment with a range of factors, including plant growth regulators (for reviews see e.g. De Jong *et al.*, 1993b; Zimmerman, 1993). However, it is clear from a limited number of studies that the formation of embryogenic cells from explants is a process in which various cell types appear in a sequential manner. This is evident from the elegant systems that allow direct embryogenesis (see e.g. Figure 11.2), where it has been shown that a number of changes occur in a cell before a division that ultimately leads to formation of a somatic embryo can take place. A particular sequence of events is also evident from studies which have monitored the fate of individual cells as they arise from an activated explant, or which have followed individual cells in established suspension cultures.

Although the growth regulator 2,4-D is commonly used to initiate somatic embryo development, it is unlikely that it has a direct and specific effect. A more likely explanation for its mode of action is that 2,4-D has a number of not necessarily very specific effects on explant cells. These effects may range from the reinitiation of cell division in previously non-dividing cells to the promotion of expansion in other cells. Only in a very small number of treated cells is a process initiated that results in the formation of embryogenic cells. This leaves the important question of why only a few cells respond by becoming embryogenic. One explanation could be that many cells can only be partially activated and are not capable of a complete change in their developmental properties. A similar phenomenon exists, for instance, in the formation of root nodules by *Rhizobia* on susceptible legume roots. While many cells in the sensitive region of the root appear to be capable of being reactivated by Nod factors produced by *Rhizobium*, only a very small number of completed root nodules are formed. It appears that only those inner cortical cells that are able to complete the cell cycle in response to Nod factors can be completely redetermined to form a root nodule. All other cells in the susceptible region of the root respond by reactivation but not by completion of the cell cycle (Yang *et al.*, 1994). An analogous situation may occur in explants activated to form embryogenic cells. While massive cell separation and proliferation occur after exposure to strong synthetic plant growth regulators, only very few truly embryogenic cells are formed. It may be that embryogenic cells are formed under the influence of totally different signal molecules, such as Nod-factor-like molecules or AGPs (reviewed by Schmidt *et al.*, 1994), perhaps produced as a secondary reaction to the treatments with growth regulators.

In this chapter a number of studies have been discussed where the question is raised as to which cellular mechanisms are involved in embryogenic cell formation. In most cases a direct correlation between the effects observed in individual cells of, for instance, an activated explant and the resulting later stages of somatic embryo formation has not been shown. Particularly in those cases where only a minority of the cells in a population are activated to develop into a somatic embryo, it is not at all clear whether the cells that form the embryo are the same as those originally assumed to be undergoing particular changes or possessing a specific morphology (Toonen *et al.*, 1994).

Acknowledgements

M.A.J.T. received support from the Technology Foundation, subsidized by The Netherlands Organization for Scientific Research.

References

Attree, S.M., Budmir, S. and Fowke, L.C. (1990) Somatic embryogenesis and plantlet regeneration from cultured shoots and cotyledons of seedlings from stored seeds of black and white spruce (*Picea mariana* and *Picea glauca*), *Can. J. Bot.* **68**, 30–34.

Backs-Hüsemann, D. and Reinert, J. (1970) Embryobildung durch isolierte Einzelzellen aus Gewebekulturne von *Daucus carota. Protoplasma* **70**, 49–60.

Becwar, M., Nagmani, R. and Wann, S.R. (1990) Initiation of embryogenic cultures and somatic embryo development in loblolly pine (*Pinus taeda*). *Can. J. For. Res.* **30**, 810–817.

Berger, F., Taylor, A. and Brownlee, C. (1994) Cell fate determination by the cell wall in early *Fucus* development. *Science* **263**, 1421–1423.

Binarova, P., Straatman, K., Hause, B., Hause, G. and Van Lammeren, A.A.M. (1993) Nuclear DNA synthesis during the induction of embryogenesis in cultured microspores and pollen of *Brassica napus* L. *Theor. Appl. Genet.* **87**, 9–16.

Börge, L., Stefanov, M., Abrahám, I., Somogyi, I. and Dudits, D. (1990) Differences in response to 2,4-dichlorophenoxy acetic acid (2,4-D) treatment between embryogenic and non-embryogenic lines of alfalfa. In: *Progress in Plant Cellular and Molecular Biology* (eds H.J.J. Nijkamp, L.H.W. Van der Plas and J. Van Aartrijk). Kluwer Academic Publishers, Dordrecht, pp. 427–436.

Bothwick, H. A. (1931) Development of the macrogametophyte and embryo of *Daucus carota*. *Bot. Gaz.* **92**, 23–44.

Botti, D. and Vasil, I. K. (1984) Ontogeny of somatic embryos of *Pennisetum americanum*. II. In cultured immature fluorescences. *Can. J. Bot.* **62**, 1629–1635.

Carpita, N.C. and Gibeaut, D.M. (1993) Structural models of primary cell walls in flowering plants — consistency of molecular structure with the physical properties of the walls during growth. *Plant J.* **3**, 1–30.

Chen, Z., Klockare, R. and Sundqvist, C. (1994) Origin of somatic embryogenesis is proliferating root primordia in seed derived oat callus. *Hereditas* **120**, 211–216.

Cordewener, J., Booij, H., Van der Zandt, H., Van Engelen, F., Van Kammen, A. and de Vries, S. (1991) Tunicamycin-inhibited carrot somatic embryogenesis can be restored by secreted cationic peroxidase isoenzymes. *Planta* **184**, 478–486.

Cosgrove, D.J. (1993) How do plant cell walls extend? *Plant Physiol.* **102**, 1–6.

Coutos-Thevenot, P., Jouanneau, J.P., Brown, S., Petiard, V. and Guern, J. (1990) Embryogenic and non-embryogenic cell lines of *Daucus carota* cloned from meristematic cell clusters: relation with cell ploidy determined by flow cytometry. *Plant Cell Rep.* **8**, 605–608.

De Jong, A.J., Heidstra, R., Spaink, H.P., Hartog, M.V., Meijuer, E.A., Hendriks, T., Loschiavo, F., Trzi, M., Bisseling, T., Van Kammen, A. and de Vries, S.C. (1993a) *Rhizobium* lipo-oligosaccharides rescue a carrot somatic embryo mutant. *Plant Cell* **5**, 615–620.

De Jong, A.J., Schmidt, D.L. and de Vries, S.C. (1993b) Early events in higher plant embryogenesis. *Plant Mol. Biol.* **22**, 367–377.

Dijak, M. and Simmons, D.H. (1988) Microtubule organization during early direct embryogenesis from mesophyll protoplasts of *Medicago sativa* L. *Plant Sci.* **58**, 183–191.

Dolezelová, M., Dolezel, J.and Nesticky, M. (1992) Relationship of embryogenic competence in maize (*Zea mays* L.) leaves to mitotic activity, cell cycle and nuclear DNA content. *Plant Cell Tissue Org. Cult.* **31**, 215–221.

Dos Santos, A.P., Cutter, E.G. and Davey, M.R. (1983) Origin and development of somatic embryos in *Medicago sativa* L. (alfalfa). *Protoplasma* **117**, 107–115.

Dubois, T., Tuedrira, M., Dubois, J. and Vasseur, J. (1991) Direct somatic embryogenesis in leaves of *Cichorium*: a histological and SEM study of early stages. *Protoplasma* **162**, 120–127.

Dudits, D., Bögre, L. and Györgyey, J. (1991) Molecular and cellular approaches to the analysis of plant embryo development from somatic cells *in vitro*. *J. Cell Sci.* **99**, 475–484.

Dudits, D., Börge, L., Bakó, L., Dedeoglu, D., Magyar, Z., Kapros, T. Felföldi, F. and Gyorgyey, J. (1993) Key components of cell cycle control during auxin-induced cell division. In: *Molecular and Cell Biology of the Plant Cell Cycle* (eds J.C. Ormrod and D. Francis). Kluwer Academic Publishers, Dordrecht, pp. 111–131.

Duijs, J.G., Voorrips, R.E., Visser, D.L. and Custers, J.B.M. (1992) Microspore culture is successful in most crop types of *Brassica oleracea* L. *Euphytica* **60**, 45–55.

Faure, O. and Nougarède, A. (1993) Nuclear DNA content of somatic and zygotic embryos of *Vites vinifera* cv. Genache noir at the torpedo stage. *Protoplasma* **176**, 145–150.

Gavish, H., Vardi, A. and Fluhr, R. (1992) Suppression of somatic embryogenesis in citrus cell cultures by extracellular proteins. *Planta* **186**, 511–517.

Guzzo, F., Baldan, B., Mariani, P., Loschiavo, F. and Terzi, M. (1994) Studies on the origin of totipotent cells in explants of *Daucus carota* L. *J. Exp. Bot.* **45**, 1427–1432.

Hilbert, J.L., Dubois, T. and Vasseur, J. (1992) Detection of embryogenesis-related proteins during somatic embryo formation in *Cichorium. Plant Physiol. Biochem.* **30**, 733–741.

Ho, W.-J. and Vasik, I.K. (1983) Somatic embryogenesis in sugarcane (*Saccharum officinarum* L.) I. The morphology and physiology of callus formation and the ontogeny of somatic embryos. *Protoplasma* **118**, 169–180.

Jones, T.J. and Rost, T.L. (1989a) The developmental anatomy and ultrastructure of somatic embryos from rice (*Oryza sativa* L.) scutellum epithelial cells. *Bot. Gaz.* **150**, 41–49.

Jones, T.J. and Rost, T.L. (1989b) Histochemistry and ultrastructure of rice (*Oryza sativa*) zygotic embryogenesis. *Am. J. Bot.* **76**, 504–520.

Kapros, T., Bögre, L., Németh, K., Bakó, L., Györgyey, J., Su, S.C. and Dudits, D. (1992) Differential expression of histone H3 gene variants during cell cycle and somatic embryogenesis in alfalfa. *Plant Physiol.* **98**, 621–625.

Karlsson, S.B. and Vasil, I.K. (1986a) Growth, cytology and flow cytometry of embryogenic cell suspension cultures of *Panicum maximum* Jacq. and *Pennisetum purpureum* Schum. *J. Plant Physiol.* **123**, 211–227.

Karlsson, S.B. and Vasil, I.K. (1986b) Morphology and ultrastructure of embryogenic cell suspension cultures of *Panicum maximum* (guinea grass) and *Pennisetum purpureum* (napier grass). *Am. J. Bot.* **73**, 894–901.

Kato, H. and Takeuchi, M. (1963) Morphogenesis *in vitro* starting from single cells of carrot root. *Plant Cell Physiol.* **4**, 243–245.

Kielisewski, M.J. and Lamport, D.T.A. (1994) Extensin: repetitive motifs, functional sites, post-translational codes and phylogeny. *Plant J.* **5**, 157–172.

Knox, J.P., Day, S. and Roberts, K. (1989) A set of surface glycoproteins forms an early marker of cell position, but not cell type, in the root apical meristem of *Daucus carota* L. *Development* **106**, 47–56.

Konar, R.N., Thomas, E. and Street, H.E. (1972) Origin and structure of embryoids arising from epidermal cells of the stem of *Ranuculus sceleratus* L. *J. Cell Sci.* **11**, 77–93.

Kreuger, M. and Van Holst, G.J. (1993) Arabinogalactan proteins are essential in somatic embryogenesis of *Daucus carota* L. *Planta* **189**, 243–248.

Kreuger, M. and Van Holst, G.J. (1995) Arabinogalactan-protein epitopes in somatic embryogenesis of *Daucus carota* L. *Planta* (in press).

Kristensen, M.M.H., Find, J.I., Floto, F., Moller, J.D., Norgaard, J.V. and Krogstrup, P. (1994) The origin and development of somatic embryos following cryopreservation of an embryogenic suspension culture of *Picea sitchensis. Protoplasma* **182**, 65–70.

Lu, C.-Y. and Vasil, I.K. (1985) Histology of somatic embryogenesis in *Panicum maximum* (guinea grass). *Am. J. Bot.* **72**, 1908–1913.

Maheswaran, G. and Williams, E.G. (1985) Origin and development of somatic embryoids formed directly on immature embryos of *Trifolium repens in vitro. Ann. Bot.* **56**, 619–630.

Mansfield, S.G. and Briarty, L.G. (1991) Early embryogenesis in *Arabidopsis thaliana*. II. The developing embryo. *Can. J. Bot.* **69**, 461–476.

Mayer, U., Buttner, G. and Jurgens, G. (1993) Apical-basal pattern formation in the *Arabidopsis* embryo — studies on the role of the *gnom* gene. *Development* **117**, 149–162.

Nagmani, R., Becwar, M.R. and Wann, S.R. (1987) Single-cell origin and development of somatic embryos in *Picea abies* (L.) Karst. (Norway spruce) and *P. glauca* (Moench) Voss (white spruce). *Plant Cell Rep.* **6**, 157–159.

Nomura, K. and Komamine, A. (1985) Identification and isolation of single cells that produce somatic embryos at a high frequency in a carrot suspension culture. *Plant Physiol.* **79**, 988–991.

Nuti Ronchi, V., Giorgetti, L., Tonelli, M. and Martini, G. (1992a) Ploidy reduction and genome segregation in cultured carrot cell lines. I. Prophase chromosome reduction. *Plant Cell Tissue Org. Cult.* **30**, 107–114.

Nuti Ronchi, V., Giorgetti, L., Tonelli, M. and Martini, G. (1992b) Ploidy reduction and genome segregation in cultured carrot cell lines. II. Somatic meiosis. *Plant Cell Tissue Org. Cult.* **30**, 115–119.

Pennell, R.I., Janniche, L., Kjellbom, P., Schofield, G.N., Peart, J.M. and Roberts, K. (1991) Developmental regulation of a plasma membrane arabinogalactan protein epitope in oilseed rape flowers. *Plant Cell* **3**, 1317–1326.

Pennell, R.I., Janniche, L., Schofield, G.N., Booij, H., de Vries. S.C. and Roberts, K. (1992) Identification of a transitional cell state in the developmental pathway to carrot somatic embryogenesis. *J. Cell Biol.* **119**, 1371–1380.

Rao, P.S. and Narayanaswami, S. (1972) Morphogenic investigations in callus cultures of *Tylophora indica. Physiol. Plant.* **27**, 271–276.

Reinert, J. (1959) Uber die Kontrolle der Morphogenese und die Induction von Adventivembryonen an Gewebeculturen aus Karoten. *Planta* **53**, 318–333.

Schüller, A., Reuther, G. and Geier,T. (1989) Somatic embryogenesis from seed explants of *Abies alba. Plant Cell Tissue Org. Cult.* **17**, 53–58.

Savoure, A., Magyar, Z., Pierre, M., Brown, S., Schultze, M., Dudits, D., Kondorosi, A. and Kondorosi, E. (1994) Activation of the cell cycle machinery and the isoflavonoid biosynthesis pathway by active *Rhizobium meliloti* Nod signal molecules in *Medicago* microcallus suspensions. *EMBO J.* **13**, 1093–1102.

Schmidt, E.D.L., de Jong, A.J. and de Vries, S.C. (1994) Signal molecules in plant embryogenesis. *Plant Mol. Biol.* **26**, 1305–1313.

Schulz, S.R. and Jensen, W. (1968) *Capsella* embryogenesis: the early embryo. *J. Ultrastruc. Res.* **22**, 376–392.

Serpe, M.D. and Nothnagel, E.A. (1994) Effects of Yariv phenylglycosides on *Rosa* cell suspensions: evidence for the involvement of arabinogalactan-proteins in cell proliferation. *Planta* **193**, 542–550.

Showalter, A.M. (1993) Structure and function of plant cell wall proteins. *Plant Cell* **5**, 9–23.

Smallwood, M., Beven, A., Donavan, N., Neill, S.J., Peart, J., Roberts, K. and Knox, J.P. (1994) Localization of cell wall proteins in relation to the developmental anatomy of the carrot root apex. *Plant J.* **5**, 237–246.

Smith, D.L. and Krikorian, A.D. (1990) Low external pH replaces 2,4-D in maintaining and multiplying 2,4-D initiated embryogenic cells of carrot. *Physiol. Plant* **80**, 329–336.

Spangenberg, G., Koop, H.-U. and Schweiger, H.-G. (1985) Different types of protoplasts from *Brassica napus* L.: analysis of conditioning effects at the single cell level. *Eur. J. Cell Biol.* **39**, 41–45.

Staxen, I., Klimaszewska, K. and Bornman, C.H. (1994) Microtubular organization in protoplasts and cells of somatic embryo-regenerating and non-regenerating cultures of *Larix. Physiol. Plant* **91**, 680–686.

Swanson, E.B., Coumans, M.P., Wu, S.C., Barsby, T.L. and Beversdorf, W.D. (1987) Efficient isolation of microspores and the production of microspore-derived embryos from *Brassica napus. Plant Cell Rep.* **6**, 94–97.

Tautorus, T.E., Fowke, L.C. and Dunstan, D.I. (1991) Somatic embryogenesis in conifers. *Can. J. Bot.* **69**, 1873–1899.

Taylor, M.G. and Vasil, I.K. (1987) Analysis of DNA size, content and cell cycle in leaves of Napier grass (*Pennisetum purpureum* Schum.). *Theor. Appl. Genet.* **74**, 681–686.

Telmer, C.A., Newcomb, W. and Simmonds, D.H. (1993) Microspore development in *Brassica napus* and the effect of high temperature on division *in vivo* and *in vitro. Protoplasma* **172**, 154–165.

Toonen, M.A.J., Hendriks, T., Schmidt, E.D.L., Verhoeven, H.A., van Kammen, A. and de Vries, S.C. (1994) Description of somatic-embryo-forming single cells in carrot suspension cultures employing video cell tracking. *Planta* **194**, 565–572.

Trigano, R.N., Gray, D.J., Conger, B.V. and McDaniel, K.J. (1989) Origin of direct somatic embryos from cultured leaf segments of *Actylis glomerata. Bot. Gaz.* **150**, 72–77.

Vasil, V. and Vasil, I.K. (1982) The ontogeny of somatic embryos of *Pennisetum americanum* (L.) K. Schum. I. In cultured immature embryos. *Bot. Gaz.* **143**, 454–465.

Vietiez, F.J., Ballester, A. and Vieitez, A.M. (1992) Somatic embryogenesis and plantlet regeneration from cell suspension cultures of *Fagus sylvatica* L. *Plant Cell Rep.* **11**, 609–613.

de Vries, S.C., Booij, H., Meyerink, P., Huisman, G., Wilde, H.D., Thomas, T.L. and Van Kammen, A. (1988) Acquisition of embryogenic potential carrot cell-suspension cultures. *Planta* **176**, 196–204.

Wernicke, W. and Brettell, R. (1980) Somatic embryogenesis from *Sorghum bicolor* leaves. *Science* **287**, 138–139.

Williams, E.G., Knox, R.B. and Kaul, V. (1984) Post-pollination callose development in ovules of *Rhododendron* and *Ledum* (Ericaceae) zygote special wall. *J. Cell Sci.* **69**, 127–135.

Yang, W.C., de Blank, C., Meskiene, I., Hirt, H., Bakker, J., van Kammen, A., Franssen, H. and Bisseling, T. (1994) *Rhizobium* Nod factors reactivate the cell cycle during infection and nodule primordium formation, but the cell cycle is only completed in primordium formation. *Plant Cell* **6**, 1415–1426.

Zársky, V., Garrido, D., Ríhová, L., Tupy, J., Vicente, O. and Heberle-Bors, E. (1992) Derepression of the cell cycle by starvation is involved in the induction of tobacco pollen embryogenesis. *Sex. Plant Reprod.* **5**, 189–194.

Zimmerman, J.L. (1993) Somatic embryogenesis — a model for early development in higher plants. *Plant Cell* **5**, 1411–1423.

The generation of a legume embryo: morphological and cellular defects in pea mutants

C.-M. Liu, S. Johnson, C.L. Hedley and T.L. Wang

12.1 Introduction

Pea (*Pisum sativum* L.) has been a favoured species for experimental biology for hundreds of years, and its study has made an enormous contribution to biochemistry and genetics. The Austrian scientist, Gregor Mendel, was the first to take advantage of its self-fertility and the availability of spontaneous mutants in his breeding experiments, from which he developed the basic principles of genetics (Mendel, 1865). Considerable molecular biological and biochemical information has accrued on pea, especially on the seed (Casey *et al.*, 1993). The specific advantages of using pea as a material for investigating the mechanism of embryo development are its large embryo, inbreeding habit and ease of crossing. The large embryo makes the screening of certain mutants easier and provides sufficient material for biochemical and molecular biological investigations, even in embryo-lethal mutants. Its inbreeding habit maintains the purity of any mutant line, reducing the complexity of the genetic background. A specific disadvantage of pea is the difficulty of transformation and thus the lack of any gene-tagging system. Although genetic transformation of pea has been reported by Davies *et al.* (1993) and others (Schroeder *et al.*, 1993), the routine acquisition of a large number of transgenic plants for tagging and antisense experiments remains a problem.

Nevertheless, studies on development in peas have been aided by the large number of mutants available (Murfet and Reid, 1993). As far as the seed is concerned, there is considerable variation in size and coloration, but there are relatively few mutants that affect development *per se*. Mendel used one character,

r (for *rugosus*), that is now known to affect the development of the embryo indirectly through its osmotic relations (Wang and Hedley, 1991), and several such *rugosus* genes have now been identified (Wang and Hedley, 1993). Seed development has been studied extensively in many legumes because of their importance with regard to storage product synthesis. Little use could be made of specific mutants in this work, however, as such mutants have not existed until recently.

12.2 Legume embryology

In almost every plant, the first division of the zygote is transverse to the long axis of the cell and is often asymmetrical, producing a large basal cell and a smaller apical cell. The embryo proper is derived mainly from the apical cell. There is considerable variation in the extent of the contribution of the basal cell to the formation of the organogenic part of the embryo, and usually it forms all or part of a suspensor. On the basis both of this variation and of the plane of the first division of the apical cell, plant embryo ontogenesis has been classified into six different types: Onograd, Asterad, Chenopodiad, Solanad, Caryophyllad and Piperad. The first five follow Maheshwari's (1950) classification, and the sixth was added by Johansen (1950) to account for those zygotes which divide longitudinally (Johri, 1984).

The embryos of the Leguminosae (Papilionaceae), in contrast to those of many families, such as the Cruciferae which possess only the Onograd form, exhibit a wide range of developmental types. The sub-family Faboideae (Papilionoideae), to which the pea belongs, displays four of the six possible types, namely Onograd, Caryophyllad, Solanad and Asterad (Prakash, 1987). Pea (*Pisum sativum* L.) itself shows embryo development of the Solanad type, as outlined below. Great diversity is also displayed with regard to the form of the suspensor (Johri *et al.*, 1992; Lersten, 1983; see Figure 12.1), perhaps the most extreme example being the relatively massive multicellular and multiseriate structure seen in *Phaseolus coccineus* (Yeung and Clutter, 1979 Figure 12.1c). Pea, in keeping with all members of the tribe Vicieae (Johri *et al.*, 1992; Lersten, 1983; see Figure 12.1a and compare with 12.2a and b), has the most uniform and distinctive suspensor (Lersten, 1983) composed of four multinucleate cells (see Figure 12.2a), two of which extend considerably during early development to move the embryo away from the micropyle and into the bulk of the endosperm. The endosperm is of the nuclear type, although it has been proposed that cellulosic walls or strands are present for a time during the early development of the embryo, to anchor the latter in place (Marinos, 1970a). Work on the legume suspensor has provided the best evidence for the role of this organ in embryo nutrition. Structural specializations, in the form of wall ingrowths, that increase the surface area for absorption have been recognized in suspensors for many years (see review by Gunning and Pate, 1974), including that of pea (Marinos, 1970a). These have been used as evidence that the cells of the suspensor help to move materials between the

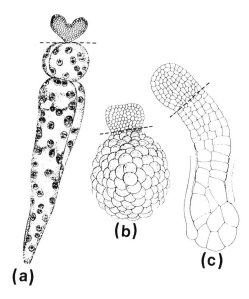

Figure 12.1. *Suspensor morphologies in the Leguminosae. (a) A member of the Vicieae,* Orobus angustifolius *(now* Lathyrus angustifolia; *Lersten, 1983). (b)* Cytisus laburnum. *(c)* Phaseolus coccineus. *The dotted line marks the margin between the embryo, above, and the suspensor. All after Guignard (1881). Not to scale.*

embryo and the seed coat and thus act as transfer cells (Gunning and Pate, 1974). More recently, direct evidence for such a role has been provided by the application of radiolabelled substances to ovules and pods of *P. coccineus* (Nagl, 1990; Yeung, 1980).

The cotyledons of legumes have been classified into four types (Smith, 1981), varying from highly leaf-like structures that expand during germination and whose function is mainly photosynthetic after germination, to those that do not show any photosynthetic function or expansion, whether they exhibit hypogeal or epigeal germination (Smith, 1981). In the latter instance, the cotyledons may turn green after emerging from the soil, but provide little photosynthate. In those mature cotyledons that have a photosynthetic function, there is some degree of differentiation into palisade and mesophyll tissues. Pea shows no clear differentiation of this type (Smith and Flinn, 1967), and the cotyledons serve only as storage organs, as in all members of the Vicieae. Pea cotyledons consist of epidermis, storage parenchyma and provascular tissues. The epidermis is a thin layer of cells, derived from the protoderm, that are much smaller than those of the underlying parenchyma. Following epidermis formation, cell divisions only occur in an anticlinal manner. Endoreduplication (endopolyploidy) occurs in the cotyledons of legumes and was present in those of all the Vicieae and about 50% of Papilionoideae examined by Smith (1981). In *Phaseolus vulgaris* and *Pisum sativum* there was a close correlation between cell and nuclear size, and DNA content (Smith, 1973, 1974).

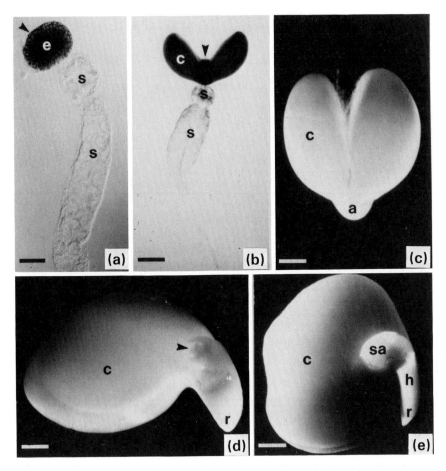

Figure 12.2. *Stages in the development of the pea embryo. (a) Late globular (stage 20); bar = 130 μm. Note formation of the shoot apical meristem (arrowed) at the top centre of the embryo. (b) Late heart-shaped (stage 22); bar = 200 μm. (c) Early cotyledon expansion; bar = 350 μm. (d) Mid-maturation (stage 23); bar = 230 μm. Note the well-developed embryonic shoot (plumule) between the cotyledons (arrowed). (e) Late maturation (stage 24); bar = 1.2 mm. All stages refer to Marinos (1970a). Embryos in (d) and (e) have had one cotyledon removed to reveal the axis. a, axis; c, cotyledon; e, embryo; h, hypocotyl; r, root; s, suspensor; sa; embryonic shoot apex.*

12.3 Embryo development in pea

The process of embryo development in dicots can be divided typically into four stages: globular, heart-shaped, torpedo-shaped and cotyledonary (see Chapter 2, Section 3, Torres Ruiz *et al.*). The embryology of pea has been studied by a number of workers (Cooper, D.C., 1938; Cooper, G.O., 1938; Marinos, 1970a; Reeve,

1948; Souèges, 1948). In this species, because of the rapid cell division and expansion that occur in the cotyledons and the delayed growth of the root axis after the heart-shaped stage, a torpedo-shaped embryo cannot be observed during embryo development. An alternative description system was proposed by Marinos (1970a) to describe the events from floral differentiation to embryo development in pea. Under his classification, the process was divided into 25 stages on the basis of a number of criteria. The first 13 stages covered flower development and fertilization, and the remainder covered seed development. During the last two stages (stages 24 and 25; Figure 12.2e) the embryo matures, accumulates storage products, amplifies its DNA, and desiccates to produce a dry seed. Hence, embryo development in pea could simply be divided into two phases, namely organ formation and organ maturation (Wang and Hedley, 1993), most of cotyledon growth being confined to the latter phase.

Following fertilization in pea, as in most plants, the zygote is divided transversely into an apical (or terminal) cell and a basal cell. The apical cell of the two-celled embryo divides transversely (c. stage 16; Liu et al., 1995; Marinos, 1970a) and the middle cell resulting from this second division plays no role in the subsequent development of the embryo proper. Thus the ontogeny of pea is of the Solanad type. As described by D.C. Cooper (1938), the basal cell divides longitudinally in the pea embryo to produce two suspensor cells, whereas the transversely divided apical cell produces an apical embryo mother cell and a middle cell. The latter, in turn, undergoes a longitudinal division to form the two-celled bulbous middle piece of the suspensor (stage 17). Several cycles of nuclear division occur in all four cells of the suspensor, but cytokinesis does not follow. As a result, the two elongated basal cells contain 64 nuclei each, and every middle cell possesses 32 nuclei (Cooper, D.C., 1938; Figure 12.2a). The suspensor is a short-lived organ which is fully developed at the proembryo stage and subsequently degenerates (after stage 22; Figure 12.2b). As mentioned above, it is believed that the suspensor channels nutrients to the embryo until the heart-shaped stage. Further nutrient transport from the mother plant to the embryo takes place via the endosperm and transfer cells (Marinos, 1970a).

Division of the apical embryonic mother cell produces an axially symmetrical globular embryo (stage 18). The protoderm in pea is not formed until the late globular stage, relatively late in comparison with cruciferous plants. In contrast, the shoot apex is initiated much earlier than in crucifers. As stated by Reeve (1948), initiation of the shoot apex occurs at about the same time as a distinct protoderm becomes established (the first visible sign of tissue differentiation) and well before the formation of the cotyledons (stages 19 to 20; Figure 12.2a). By the time the cotyledonary lobes are evident, two tunica layers are present in the shoot apex, and within them occasional periclinal divisions can be observed (Reeve, 1948). In contrast, in the crucifer *Arabidopsis thaliana*, the formation of the shoot apical dome can be delayed until the time of seed germination.

Establishment of bilateral symmetry, by the initiation of two cotyledons to form a heart-shaped embryo, is often considered to be a landmark for the end of the proembryo stage (stage 20). This process is believed to be dependent on polar

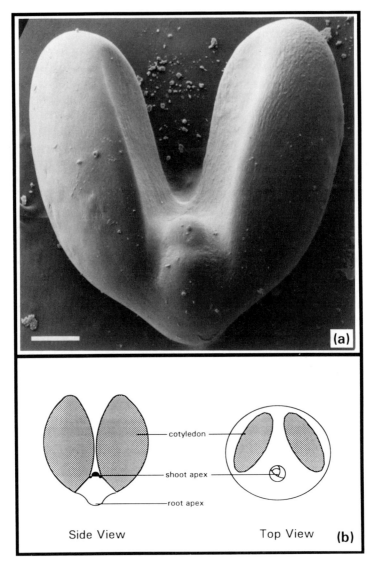

Figure 12.3. *Symmetry in the pea embryo. (a) Scanning electron micrograph of a heart-shaped pea embryo; bar = 200 μm. Note that the position of the shoot apical dome to the front of the image between the two cotyledons creates a single longitudinal plane of symmetry. Note also the lip to the rear of the apex which means that the apex can only be seen from this side.*
(b) Diagrammatic representation of the embryo in (a) (side view) and projection on to a globular embryo of the position of the progenitor meristems (top view) generating the embryo in (a).

transport of auxin, at least in *A. thaliana* and *Brassica juncea* (Liu *et al.*, 1993). According to Reeve (1948), cotyledon growth is initiated by short, curving tiers of cells called rib-meristems. As a consequence of the rapid growth of the cotyledons by cell division in these meristems, and the relatively delayed elongation of the axis, there is no real torpedo-shaped stage during embryo development in pea, as mentioned earlier. Another feature of the pea embryo that has not been reported by previous workers is that it is bilaterally symmetrical, with only one plane running longitudinally through the axis. In *B. juncea* and *A. thaliana*, there are two planes of symmetry. To clarify this point, Figure 12.3 shows a scanning electron micrograph (Figure 12.3a) and corresponding diagrams (Figure 12.3b) of a young pea embryo with its shoot apex towards the front of the image. The shoot apex faces the testa *in vivo* and it can be observed only from one side of the embryo at this stage, as is evident from the micrograph. This positioning generates only one plane of symmetry in the embryo.

The cotyledons enlarge significantly prior to any appreciable elongation of the embryo axis. Active mitotic division in the distal region and rib-meristems (see Figure 12.3b) in the basal portions contribute to the rapid growth of the cotyledons. This growth, and the delayed elongation of the axis, result in a small temporary depression at the radicle end. At least two tiers of cells are clearly evident in this depression, and these short tiers are the first indication of a columella that later contributes to the formation of the root apex (Reeve, 1948). Extension of the radicle below the bases of the cotyledons occurs only when the shoot apex has enlarged to form a high and rounded dome (stage 22; Figure 12.2b).

12.3.1 *Cellular development in cotyledons*

The cotyledons of pea, like those of all members of the Vicieae, are storage organs that accumulate large amounts of starch and protein, and a small amount of lipid for use in seed germination. As mentioned above, the cotyledons are composed of storage parenchyma, epidermis and provascular traces. The last do not develop into a functional vasculature until germination has been initiated, at which point the procambium tissue gives rise to mature xylem and phloem elements within the first 2 days (Smith and Flinn, 1967). As mentioned earlier (Section 12.2.), no specialization occurs in the organs with regard to photosynthesis. The cells of the parenchyma differ greatly in size throughout development (Ambrose *et al.*, 1987), and genotype differences can be detected (Wang and Hedley, 1993). Smith and Flinn (1967) recognized distinct zones in the parenchyma of the mature cotyledon comprising hypodermis, and inner and outer storage parenchyma, although these differences are certainly not apparent earlier in development. Throughout development, there is a shift towards a population of large cells (Ambrose *et al.*, 1987). Nuclear endoreduplication occurs in the cotyledonary parenchyma during the later stages of embryo development. As reported by Scharpé and van Parijs (1973) and subsequently confirmed by the work of Corke *et al.* (1987), the cells of the developing cotyledons continue to duplicate DNA after the cell number has reached a plateau, and the cells of fully grown cotyledons have DNA levels averaging between 32C and 64C. When the level of DNA is quantified in individual cells, endoreduplication can be detected in a few cells of very young cotyledons (shortly after stage 22). The number of cells with ploidy levels greater than the diploid level increases as the embryo grows (Corke *et al.*, 1987).

The patterns of cell division and endoreduplication within the pea cotyledons have been related to the accumulation of storage products (Corke *et al.*, 1987, 1990a,b; Hauxwell *et al.*, 1990, 1993; see Section 12.5). Storage protein accumulation can be observed only in cells with a DNA level equivalent to, or above, 5C (Corke *et al.*, 1987). Evidence based on the buoyant density and reassociation kinetics of soybean DNA has eliminated the possibility of a selected amplification of certain gene sequences during endoreduplication (Goldberg *et al.*, 1981).

12.4 Pea embryo mutants

During a mutation programme to isolate genes affecting seed development and especially storage product composition (Wang *et al.*, 1990), a population of mutagenized seed was screened for embryo mutants by opening pods during development and observing the embryos. Nine embryo mutants were isolated and have been purified to near-isogenicity by selection from plants heterozygous for the mutation. The morphological characterization of these mutants has been described in detail elsewhere (Johnson *et al.*, 1994). A separate mutation programme (Weller, unpublished data) led to the isolation of another type of mutant — one that influences the later stages of development.

12.4.1 *Cotyledon mutants*

All the mutants in this category have cotyledons that appear to be single or 'fused'. (For a more detailed consideration of this terminology, see Johnson *et al.*, 1994.) The embryos of E1137 and E2391 are similar to the wild-type embryo, except that the two cotyledons are joined together along the margin distal to the axis (see Figure 12.4). The shoot apex and root are in the same position as in the wild-type embryo. In E1137, however, there is an incomplete coalescence of the two cotyledons, as there is a depression over the top of the single cotyledon. In E2391 there is no depression. Neither mutant is lethal, and the seeds develop to maturity and can germinate to produce normal plants.

A possible explanation for the phenotypes of these two mutants might involve the positioning of the initial meristems which generate the cotyledons. In Figure 12.3b, the top view shows the position of the cotyledons with respect to the plumule. If the development of the cotyledons is a reflection of the growth in the underlying rib-meristems, to use the concept of Reeve (1948), then a simple extension of the area projected by these meristems could create a single cotyledon as shown in Figure 12.4. The degree of overlap between the meristems would dictate the completeness of the single cotyledon, and would be represented by

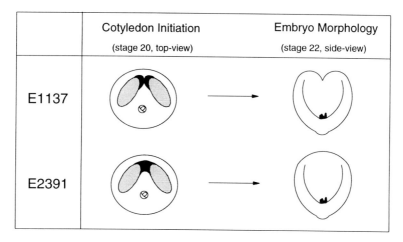

	Cotyledon Initiation (stage 20, top-view)	Embryo Morphology (stage 22, side-view)
E1137		
E2391		

Figure 12.4. *Hypothetical generation of two single cotyledon pea mutants. The left of the diagram represents the positioning of progenitor meristems as in Figure 12.3b for mutants E1137 and E2391, showing additional meristematic regions (black) required to generate the morphologies on the right. The morphologies are diagrammatic representations of the mutant embryos (Johnson et al., 1994) at the heart-shaped stage equivalent to the wild type in Figure 12.3a. Stages are according to Marinos (1970a).*

the phenotypes of E1137 and E2391; a complete overlap would create the latter phenotype.

A similar explanation can be extended to the phenotypes of the other cotyledon mutants, for example E1881. The phenotype of this mutant is like a 'spinning top', as both the shoot apex and the root are situated in the middle and opposite sides of the rather flat single cotyledon (Johnson *et al.*, 1994). Such a phenotype might be created if there was partial fusion of both rib meristems at their tips, and a displacement of them further away from the apex. Two depressions will be formed where both meristems join, and growth would be more transverse than longitudinal. Another shift in the rib meristem, this time longitudinally, to initiate cotyledons from a meristem deeper in the globular embryo and beneath the shoot meristem, could create the phenotype of mutant E4650 (Johnson *et al.*, 1994), where the shoot and root apices are forced apart. In this mutant, the embryos have a normally differentiated root, but a single, narrow cotyledon. The shoot apex is situated at the top in its correct position, although it is further away from the root compared with the apex in the other cotyledon mutants described above. The mutant embryos can survive to form seeds, but these seeds cannot develop properly upon germination, and form stunted plants. Longitudinal sections through the embryo indicate that there is no proper vascular connection between the shoot apex and the root, the cells being rather irregular, due to the displacement of cotyledon formation.

12.4.2 *Cellular mutants*

The two mutants in this group both show alterations to the normal cell shape in the embryo, one affecting the epidermis and adjacent cell layers and the other mainly affecting the vacuolate cells of the parenchyma. To date, neither phenotype has been fully reported in the literature for other species, although similar mutants appear to exist in *A. thaliana*.

The first mutant in this category, E2748, shows a defect in epidermis formation, is embryo-lethal, and thus is similar to *knolle* or *keule* (Mayer *et al.*, 1991). Its embryo has a rough irregular surface and appears a much darker green than similar sized wild-type embryos (Figure 12.5a). Although the mutant embryo has a distorted morphology, the cotyledons, shoot apex and root are still recognizable. The mutant embryos never grow sufficiently to occupy the endosperm cavity, and they abort before maturation. In the mutant embryos, instead of a layer of small cells, the outermost cells are enlarged and rounded, with very large vacuoles (Figure 12.5b). Both anticlinal and periclinal division can be seen in this layer of enlarged cells. At the late stage of embryo development, the mutant embryo literally becomes stuck to the inner surface of the testa, and it is not possible to remove it without causing damage to the cells that are in contact with each other. In sections, it is difficult to distinguish a clear boundary between the embryo and the testa tissues late in embryogenesis. The phenotype of this mutant may arise because the lack of an epidermis releases a physical restraint on the parenchyma, so that the outer cells enlarge and the correct interactions (e.g. development of transfer cells) between testa and embryo cannot occur. Alternatively, a direct alteration of the epidermal cells may have occurred, causing an increase in expansion and a change in cell division activity. The epidermis is one of the three major tissues of the radial pattern in *A. thaliana* (Mayer *et al.*, 1991), and it is one of the first signs of tissue differentiation during embryogenesis, as stated above. It has been suggested that the *knolle* mutant of *A. thaliana*, which lacks an epidermis, and the epidermis-defective mutant of carrot, *ts11*, neither of which shows the correct morphogenesis, indicate that protoderm factors are required for normal morphogenesis (Yadegari *et al.*, 1994). E2748 however, does show the correct morphogenesis in that it generates an axis and cotyledons. This may mean that E2748 does not lack a functional epidermis, or that pea does not require protoderm factors for normal morphogenesis. This type of mutant, therefore, should prove very useful for examining early differentiation in embryos and embryo-mother plant interactions.

The second mutant in this category, E1735, *cyd*, is a mutant that displays defective cytokinesis (Johnson *et al.*, 1994; Liu *et al.*, 1995). Although *cyd* embryos are much smaller than those of the wild type, like E2748, they have basic tissue patterns. The embryos have an uneven surface and regions of translucency (Figure 12.5c). Sections through the *cyd* embryo show that the cell morphology and arrangement have changed greatly, especially in the cotyledons. Most cotyledonary cells in *cyd* mutants are round, multinucleate and contain partially formed cell walls (Figure 12.5d). A detailed description of the mutant

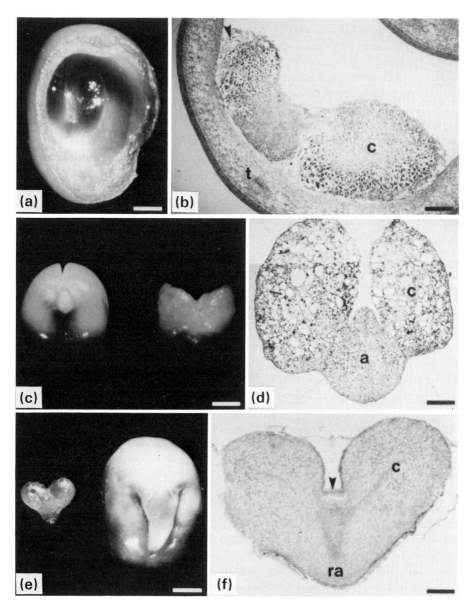

Figure 12.5. *Pea embryo mutant phenotypes. E2748: (a) The embryo within its testa; bar = 1.1 mm; (b) Transverse section through the embryo and testa, the arrowhead indicating enlarged cells at the periphery of the cotyledon; bar = 111 μm. E1735, cyd: (c) A wild-type embryo (left) from the same pod as the E1735 mutant embryo (right); bar = 1.7 mm; (d) Longitudinal section through the embryo; bar = 160 μm. E6101: (e) An E6101 mutant embryo (left) from the same pod as the wild-type (right); bar = 1.7 mm; (f) Longitudinal section through an E6101 embryo, the arrowhead indicating the position of the shoot meristem; bar = 187 μm. a, axis; c, cotyledon; ra, root axis; t, testa.*

phenotype can be found elsewhere (Johnson *et al.*, 1994; Liu *et al.*, 1995). To date, the smallest embryos in which it is possible to identify the cell wall defect are at the early-heart stage. An embryo mutant of *A. thaliana*, *emb101* (Castle *et al.*, 1993), shows some features (e.g. large cell size) displayed by *cyd* embryos (D. Meinke, personal communication).

The fact that cell plates are formed normally in meristem cells but are disrupted in the vacuolate cells, producing partially formed cell plates, or stubs, indicates that the initiation of cytokinesis is not interrupted by the *cyd* mutation. The major defect in *cyd* embryos concerns the completion of the cell plate to conclude cytokinesis. Late septation mutants of yeast are phenotypically similar to *cyd*, in that they have giant multinucleate cells (Fankhauser and Simanis, 1994), although the mechanism of cytokinesis in yeast may be quite different to that in higher plants. It has been suggested that the defect in *cyd* may be associated with the cytoskeletal apparatus or with the vesicle-trafficking system of the cell (Liu *et al.*, 1995).

The multinucleate nature of the cells of *cyd* cotyledons becomes clear when individual cells are examined by fluorescence microscopy. In cotyledons of a 29-mg embryo, 72% of the cells have two or more nuclei, most of them have 2 to 16 nuclei, and the maximum number of nuclei in a single cell of an embryo of this weight was 40. In a 40-mg E1735 embryo there is no significant difference in the number of cotyledon cells with more than two nuclei, but the maximum number of nuclei in a single cell increases to 118. In some nuclei, *c.* 1000 chromosomes can be observed at metaphase. These nuclei may have been formed from the fusion of several nuclei during prophase when the nuclear membranes were broken down. The large numbers of nuclei with high ploidies mean that many cells will contain very high levels of DNA. Quantification of the DNA content of individual cells, using image analysis of DAPI-stained nuclei, confirms that the mutant does indeed contain cells with very large amounts of DNA (almost 1000C in some cells) compared with the wild type (Figure 12.6). The high levels of DNA indicate that some cells must go through up to nine karyokinesis cycles without completing cytokinesis.

The effect of the *cyd* gene product is not embryo-specific (Liu *et al.*, 1995). Embryos of the mutant normally die during the later stages of seed development. It is not thought that the mutation is embryo-lethal *per se*, but rather that the embryo becomes stranded when the endosperm disappears, since it does not grow sufficiently to fill the embryo sac. If *cyd* embryos are removed before the liquid endosperm disappears, seedlings can be recovered by inducing precocious germination using the method of Cook *et al.*, 1988. The *cyd* seedlings are not normal (Figure 12.7a), they have short internodes and unexpanded leaves, and they die when transferred to soil. When shoot apices are excised after the first five leaves have been produced (as several nodes already exist in the mature embryo) and examined in section, a normal tissue structure and cell arrangement is observed. Cell differentiation is also quite normal in the mutant, but a few cells with cell-wall stubs can be observed (Figure 12.7b), indicating that the defect in cytokinesis is not embryo-specific. The stubs occur in cells of the leaf and stem,

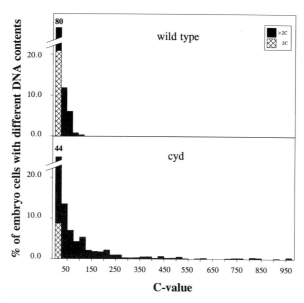

Figure 12.6. *DNA levels in cells of wild-type and* cyd *embryos. Cells were separated by acid hydrolysis (Liu* et al., *1995), stained with DAPI and viewed using fluorescence microscopy. DNA levels were quantified using image analysis (Seescan, Cambridge, UK) of total fluorescence by reference to root-tip standards. The normal diploid level of DNA, 2C, is represented by the hatched column within the first group (1C–25C).*

but never in cells of the meristem. The frequency of abnormal cell-plate formation is also much lower than that in the cotyledons. It has been suggested (Liu *et al.*, 1995) that the low frequency could be due to the fact that, in the seedlings, there were very few opportunities for vacuolated cells to divide. In addition, when callus is produced by aseptically culturing apical tissue in a hormone-supplemented medium, the resulting tissue contains a high frequency of cells with cell-wall stubs, even after several subcultures (Figure 12.7c). The occurrence of the phenotype in callus tissues supports the hypothesis that vacuolation is required, since it is a prominent part of the growth cycle of such tissues.

12.4.3 *Developmentally blocked mutants*

There are two pea mutants in which embryo morphogenesis is blocked at the heart-shaped stage (stage 22; Figure 12.2b), namely E6101 and E1450. The embryos of these two mutants have two small cotyledons and an unelongated root. Morphogenesis is blocked at slightly different stages in the two mutants, with E1450 embryos apparently arrested at a slightly earlier stage than those of E6101. The latter is shown in Figure 12.5e. Shoot apices are not fully formed in either mutant, but the root meristems appear to be normal. In E1450, the organ at the position of the apex expands considerably to form a bulbous structure with

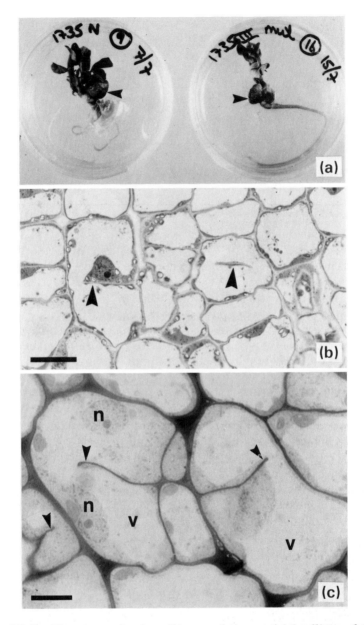

Figure 12.7. *Phenotype of* cyd *seedlings and tissues. (a) Seedlings of the wild type (left) and E1735,* cyd *(right) from precociously germinated embryos in tissue culture. Note that no lateral roots are produced by the* cyd *seedling. The cotyledons are indicated by arrowheads. (b) Londitudinal section of a young stem from a* cyd *seedling showing cell-wall stubs (arrowheads) following precocious germination in vitro; bar = 20 μm. (c) A section of callus generated from a* cyd *seedling showing cell-wall stubs (arrowheads); bar = 10 μm. n, nucleus; v, vacuole.*

highly vacuolated cells which shows no resemblance to a shoot apex with regard to tissue organization (Johnson et al., 1994).

12.4.4 *Viviparous mutants*

Viviparous mutants have been isolated from several plant species, maize probably being the best known (see Chapter 6, Clark). Numerous mutants which lack normal dormancy have also been isolated (e.g. in *A. thaliana*; Karssen and van Loon, 1992). In many of these mutants the plant hormone abscisic acid (ABA) has been implicated in the generation of the phenotype. Peas do not show true dormancy, and investigations into the involvement of ABA in seed development and precocious germination *in vitro* have left its role equivocal (e.g. Barratt et al., 1989). A mutant that may help to resolve this problem in pea has been isolated recently. This mutant affects the maturation stage of embryogenesis in pea.

Following an EMS mutagenesis programme of cv. Torsdag (conducted by J. Weller, University of Tasmania), a mutant was recovered that germinates viviparously in the pod 17–25 days post-anthesis (Figure 12.8). The seeds are viable if removed from the pod and planted, but die during normal seed desiccation if they remain on the mother plant (Presser and Reid, unpublished data). As a result of experiments designed to analyse (a) the amount of ABA in the mutants compared with the wild type, (b) the sensitivity of isolated embryos to ABA in culture and (c) the production of ABA under conditions of water stress,

Figure 12.8. *A viviparous mutant of pea germinating in the pod* c. *17 days post-anthesis (courtesy J. Presser).*

it has been concluded that the mutant phenotype is not related to a defect in ABA metabolism but to a deficiency in another factor that prevents precocious germination (Presser and Reid, personal communication). Furthermore, the phenotype was found to be unrelated to the expression of the pea homologue of the *A. thaliana abi3* gene which affects ABA sensitivity in the seed (Presser and Reid, personal communication). An investigation however, of the expression of the late embryogenesis-abundant protein PsB12, a pea dehydrin, has shown that PsB12 gene expression is correlated with germination in the mutant, in that the level of its mRNA decreases dramatically as precocious germination progresses (Presser and Reid, unpublished data).

12.5 Storage protein synthesis and development

Seed proteins in pea can be divided into two categories: the albumins and the globulins (Casey *et al.*, 1993). The former include small amounts of numerous 'housekeeping' proteins, so called because they are essential for normal cell metabolism, whereas the latter consist of the true storage proteins, namely vicilin, legumin and convicilin (Casey *et al.*, 1993). A seed storage protein is normally defined as any protein that is accumulated in significant quantities in the developing seed and which is rapidly hydrolysed on germination to provide a reduced nitrogen source for the early stages of seedling growth.

In pea embryos, the main storage proteins are vicilin and legumin (Domoney and Casey, 1985). The accumulation of such proteins provides a way of studying embryo development, as the proteins are considered to be useful markers of development (Heath *et al.*, 1986). The expression of storage protein genes and the distribution of the proteins are regulated both temporally and spatially. Their initiation occurs relatively early in the cotyledonary expansion phase when there is an increase in fresh weight (Corke *et al.*, 1987; Dure, 1975). Messenger RNA *in situ* hybridization studies have revealed patterns of gene expression which have indicated a distinct temporal and spatial pattern (Harris *et al.*, 1989; Hauxwell *et al.*, 1990). Vicilin mRNA is detectable first, being localized in the parenchyma cells of the adaxial region of the cotyledon closest to the young shoot apex (Figure 12.9a). The mRNA distribution then increases by spreading down and across in a wave-like manner as cotyledon growth occurs (Hauxwell *et al.*, 1990, 1993; Wang and Hedley, 1993; Figure 12.9b). Legumin gene expression follows a similar pattern, but occurs a little later than that of vicilin as determined on a fresh weight basis (Hauxwell *et al.*, 1993; Wang and Hedley, 1993). Certain areas of the embryo, namely the provascular elements, epidermis and embryonic axis, do not appear to express any storage protein genes, which may be due to the fact that these areas continue to divide actively until late in development (Harris *et al.*, 1989).

The link between the switch from cell division to cell expansion and subsequent storage protein deposition has been studied by the manipulation of embryo growth using tissue culture (Corke *et al.*, 1990a). Embryos grown *in vitro* appear morphologically similar to those grown *in vivo*, with embryos below 5 mg fresh

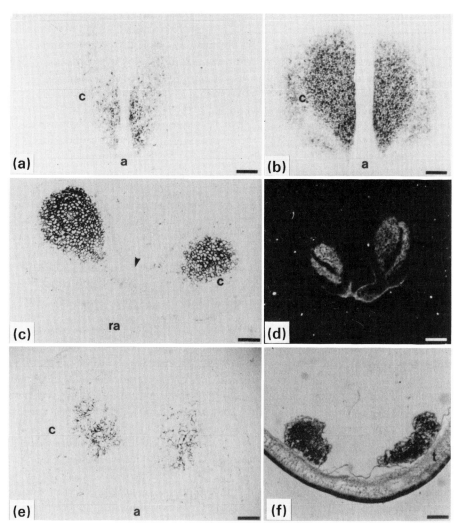

Figure 12.9. *Localization of storage protein and its mRNA in wild-type pea and embryo mutants. (a) 25-mg fresh weight wild type; bar = 214 μm. (b) 50-mg wild type; bar = 180 μm. (c) E6101, the arrowhead indicating the position where the shoot meristem should be located; bar = 113 μm. (d) E6101, note the clear strand towards the centre of the cotyledon at the position of the provascular tissue; bar = 227 μm. (e) E1735, bar = 250 μm. (f) E2748; bar = 114 μm. In situ hybridizations to mRNA (a, b, c, e and f) were carried out using digoxygenin-labelled cRNAs and developed using an anti-digoxygenin antibody coupled to alkaline phosphatase. Storage protein deposition was detected using an affinity-purified anti-vicilin antibody (a gift from C. Domoney and R. Casey) and detected using a gold-labelled secondary antibody which was then silver enhanced. a, axis; c, cotyledon; ra, root axis; t, testa.*

weight showing the greatest amount of relative growth. Some cells begin to expand and cease to divide almost immediately after the embryos are placed in culture, and after 60 h all cell division ceases. This confirms previous reports (Ambrose *et al.*, 1987) that there is a decrease in cell number and an increase in cell volume in embryos in culture compared to embryos of an equivalent weight *in vivo*. The outcome of the change in cell division activity explains an earlier observation (Domoney *et al.*, 1980) that embryos *in vitro* have higher levels of storage protein. It would seem that storage protein synthesis occurs when cell division ceases and cell expansion begins. This is consistent with the observation that tissues undergoing division — the provascular elements, epidermis and embryonic axes — do not contain the proteins or express the genes. Yang *et al.* (1990) examined the expression of storage protein genes in normal embryos compared to those grown in culture using cDNA probes for the major storage protein classes, vicilin, legumin and convicilin. All the classes are expressed in embryos *in vitro* but, in many instances, at a greater level than *in vivo*. These differences parallel the shift in the cotyledonary cell population from cell division to cell expansion *in vitro* in that they are greatest in young embryos (1–10 mg). Such embryos are more active in division and have the greatest capacity for expansion. Hence *in vitro* culture brings about the premature ageing of the embryos, since they naturally show the trend towards cell expansion during normal development.

The growth of embryos *in vitro* has also been manipulated by the use of xenobiotic compounds (Corke *et al.*, 1990b) with aphidicolin yielding the most interesting results. This compound limits DNA levels in pea cotyledon cells by blocking DNA synthesis and thus endoreduplication. Its application to developing embryos indicates that storage protein deposition is not dependent on a certain DNA level being attained, and that endoreduplication and storage protein deposition are concomitant rather than dependent events. Since growth of the embryos continues in the absence of DNA synthesis, the observed increase in weight must be due to cell expansion; storage protein deposition is enhanced under these conditions.

The manipulation of cellular development in pea can also be achieved by generating embryo mutants as outlined above, without resorting to *in vitro* and chemical treatments. We have started to analyse storage protein production in some of the pea embryo mutants in order to investigate further those features of cell division, cell expansion and DNA endoreduplication that are commonplace in legume embryos, especially embryos of the Vicieae. The mutants which are described fully above (Section 12.3) and which will be discussed further here are those with altered cell types and patterns, namely E2748 and E1735, respectively, and one mutant that is blocked in its development, namely E6101.

E6101 embryos (see Section 12.4.3) reach only 5% of the weight of the wild type and have very rounded cotyledons (Figure 12.5e). The cells show little or no expansion (Figure 12.5f). When mRNA *in situ* hybridizations are carried out with both vicilin and legumin cRNA probes, the tissues are heavily labelled, much more so than would be expected for an embryo of this fresh weight and apparent developmental stage (Figure 12.9c). Furthermore, the spatial pattern of expression is markedly different to that of the wild type (Hauxwell *et al.*, 1990; Figures 12.9a and b). The labelling is distinctly more central, with a zone of several layers of cells beneath the adaxial epidermis that is free of labelling. This contrasts with the wild type in that the labelling occurs in those cells immediately adjacent to the epidermis. Whether this is due to extra cell divisions occurring in this region, or whether it represents a spatial shift in the population of cells expressing vicilin genes, is uncertain at present. Immunolabelling with an anti-vicilin antibody shows that the protein and its mRNA are co-localized (Figure 12.9d). These results indicate that storage protein expression may not be linked to the onset of cell expansion, since these embryos have much smaller cells than the wild type. The expression may be activated, therefore, by the cessation of cell division alone. Further research is needed, however, to determine the number of cells, their size distribution, and the mitotic regions in the embryos of E6101, before any definite conclusions can be drawn.

When *in situ* mRNA hybridizations are carried out on sections from E1735, *cyd*, embryos (see Section 12.4.2), the mutant accumulates vicilin mRNA (Figure 12.9e) and, as in the case of E6101, the corresponding protein (data not shown). The large highly vacuolated cells in this mutant make interpretation of the signal strength and exact spatial expression of genes a little difficult, but it seems that there is less vicilin mRNA than would be expected. The pattern of expression, however, is only slightly altered by comparison with the wild type, the earliest expression being observed in the adaxial region close to the shoot apex. Hence this mutant, despite the disruption to the division process and retention of nuclear division, still produces storage protein.

As mentioned earlier (see Section 12.4.2), E2748 appears to lack a proper epidermis and is in very close contact with the inside of the testa wall (Figure 12.5b) via a layer of highly vacuolated cells. It is not known whether these cells originate from the parenchyma or the epidermis. The cotyledons do not mature fully, but are not blocked in development as are those of E6101. Again, *in situ* mRNA hybridizations reveal that these embryos contain a large amount of vicilin mRNA throughout their cotyledons (Figure 12.9f) . As with the other mutants, immunolabelling shows co-localization of the protein and mRNA (data not shown).

Storage protein production has also been used to study embryo development in *A. thaliana*. As many *A. thaliana* embryo mutants are lethal (see Chapter 3, Section 2, Meinke), the stage at which they abort or at which their development is blocked can be determined by examining the occurrence and levels of storage protein production (Heath *et al.*, 1986; Patton and Meinke, 1990; Schwartz *et al.*, 1994). In their initial studies, Meinke and co-workers (Heath *et al.*, 1986) concluded that there was a good correlation between developmental arrest and inability to accumulate storage proteins; most of the mutants blocked in morphogenesis were also blocked in cell differentiation and did not accumulate the proteins, although they remained viable through to the final stages of maturation. However, later analyses of *A. thaliana* mutants (Patton and Meinke, 1990), including *sus* (Schwartz *et al.*, 1994; see also Chapter 3, Section 4.4) and *raspberry* (Yadegari *et al.*, 1994) mutants, showed that storage protein and starch synthesis could occur in the absence of the correct morphogenetic changes. The *raspberry* and *sus* mutants also produced storage protein mRNA in their suspensors. Thus morphogenesis and cell differentiation can be uncoupled during the development of the *A. thaliana* embryo.

12.6 Conclusions

The conclusion from our studies of pea is essentially the same as that for *A. thaliana*, since it is clear that the pea storage proteins can be produced in the absence of correct development. To date, therefore, the only consistent parameter that correlates with the initiation of storage protein gene expression in pea is the cessation of cell division. The uncoupling of morphogenesis and cell differentiation in embryo mutants poses the question of how cells know when to activate transcription and produce storage protein. Do they contain positional and chronological information that dictates whether they will produce storage protein? Can they measure time? Three possible mechanisms have been suggested to account for the temporal and spatial regulation of storage protein production in soybean (Perez-Grau and Goldberg, 1989), namely cell-to-cell signals, a gradient of regulatory substances, or an intrinsic clock mechanism. It appears that storage protein synthesis does not occur in cells that are actively dividing, and teleologically this makes sense as both processes have a high demand for energy and metabolites, and would be in competition with each other for such resources. In wild-type pea cotyledons, the cells that accumulate storage protein first are the 'oldest' cells located at the adaxial side; they are the largest cells and the ones in which cell division ceases first (Smith, 1973). Manipulation of cellular development *in vitro* prematurely ages cells and increases their storage protein content (Yang *et al.*, 1990). It has been suggested, therefore, that the cells know their identity, can measure time, and that the accumulation of storage protein is an intrinsic function. The simplest way for these cells to measure time is for them to count the number of cell or endoreduplication cycles through which they have passed (Hauxwell *et al.*, 1993). It would be extremely interesting, therefore, to follow cell cycle progression in the cotyledonary cells of the new mutants in order to determine whether it is altered.

Although the molecular analysis of embryo mutants in pea is difficult in the absence of tagging and map-based cloning techniques, technologies such as differential screening and mRNA differential display can be applied readily (see also Chapter 2, Section 8.2, Torres-Ruiz *et al.*). The availability of near-isogenic lines greatly improves the possibility of discovering genes that are expressed differently as a consequence of the mutation of interest. Early results using differential display with the *cyd* isolines have been promising, and a gene with homologies to a ribosomal protein has been cloned (Liu and Wang, unpublished data). The gene is highly overexpressed in *cyd* embryos compared with those of the wild type, indicating that gene suppression (see also Chapter 3, Section 5.1, Meinke) may be important in the generation of a normal phenotype. In addition to the information that has already been obtained regarding the interplay between cell differentiation and morphogenesis, such molecular analyses, coupled to current cell biological investigations of the pea mutants, should provide insights into the mechanisms underlying embryogenesis, and especially those characteristics that are either prevalent in or peculiar to legumes.

Acknowledgements

The John Innes Centre is supported by a grant-in-aid from the BBSRC. This work was also supported by MAFF. The authors would like to thank Jenny Presser and Jim Reid, University of Tasmania, for providing unpublished material, Rod Casey and Claire Domoney, JIC, for the provision of storage protein antibodies and cDNAs and Kim Findlay, JIC for help with scanning electron microscopy. The PsB12 probe was kindly provided for J. Presser and J. Reid by Dr M. Robertson, CSIRO, Canberra.

References

Ambrose, M.J., Wang, T.L., Cook, S.K. and Hedley, C.L. (1987) An analysis of seed development in *Pisum sativum*. IV. Cotyledon cell populations *in vivo* and *in vitro*. *J. of Exp. Bot.* **38**, 1909-1920.

Barratt, D.H.P., Whitford, P.N., Cook, S.K., Butcher, G. and Wang, T.L. (1989) An analysis of seed development in *Pisum sativum*. VIII. Does abscisic acid prevent precocious germination and control storage protein synthesis? *J. of Exp. Bot.* **40**, 1009-1014.

Casey, R., Domoney, C. and Smith, A.M. (1993) Biochemistry and molecular biology of seed products. In: *Peas: Genetics, Molecular Biology and Biotechnology* (eds R. Casey and D.R. Davies). CAB International, Wallingford, pp. 121-163.

Castle, L.A., Errampalli, D., Atherton, T.L., Franzmann, L.H., Yoon, E.S. and Meinke, D.W. (1993) Genetic and molecular characterization of embryonic mutants identified following seed transformation in *Arabidopsis*. *Mol. and Gen. Genet.* **241**, 504-514.

Cook, S.K., Adams, H., Hedley, C.L., Ambrose, M.J. and Wang, T.L. (1988) An analysis of seed development in *Pisum sativum*. VII. Embryo development and precocious germination *in vitro*. *Plant Cell, Tissue Organ Culture* **14**, 89-101.

Cooper, D.C. (1938). Embryology of *Pisum sativum*. *Bot. Gazette* **100**, 123-132.

Cooper, G.O. (1938). Cytological investigations of *Pisum sativum*. *Bot. Gazette* **99**, 584-591.

Corke, F.M.K., Hedley, C.L., Shaw, P.J. and Wang, T.L. 1987. An analysis of seed development in *Pisum sativum*. V. Fluorescence triple staining for investigating cotyledon cell development. *Protoplasma* **140**, 164-172.

Corke, F.M.K., Hedley, C.L. and Wang, T.L. (1990a). An analysis of seed development in *Pisum sativum*. Cellular development and the deposition of storage protein in immature embryos grown *in vivo* and *in vitro*. *Protoplasma* **155**, 127-135

Corke, F.M.K., Hedley, C.L. and Wang, T.L. (1990b) An analysis of seed development in *Pisum sativum*. *In vitro* manipulation of embryo development using xenobiotic compounds. *Protoplasma* **155**, 136-143.

Davies, D.R., Hamilton, J. and Mullineaux, P. (1993) Transformation of peas. *Plant Cell Rep.* **12**, 180-183.

Domoney, C. and Casey, R. (1985) Measurement of gene number for seed storage proteins in *Pisum*. *Nucleic Acids Res.* **13**, 687-699.

Domoney, C., Davies, D.R. and Casey, R. (1980) The initiation of legumin synthesis in immature embryos of *Pisum sativum* L. grown *in vivo* and *in vitro*. *Planta* **149**, 454–460.

Dure, L.S. III. (1975) Seed formation. *Ann. Rev. Plant Physiol.* **26**, 259–278.

Fankhauser, C. and Simanis, V. (1994) Cold fission: splitting the *pombe* cell at room temperature. *Trends Cell Biol.* **4**, 96–101.

Goldberg, R.B., Hoschek, G., Ditta, G.S. and Breidenbach, R.W. (1981) Developmental regulation of cloned superabundant embryo mRNAs in soybean. *Dev. Biol.* **83**, 218–231.

Guignard, M.L. (1881) Recherches d'embryogénie végétale comparée. Premier mémoire: légumineuses. *Ann. Sciences Naturelles Botanique, Series VI* **12**, 5–166.

Gunning, B.E.S. and Pate, J.S. (1974) Transfer cells. In: *Dynamic Aspects of Plant Ultrastructure* (ed. A.W. Robards). McGraw-Hill, London, pp. 441–480.

Harris, N., Grindley, H., Mulchrone, J. and Croy, R.R.D. (1989) Correlated *in situ* hybridization and immunochemical studies of legumin storage protein deposition in pea (*Pisum sativum* L.). *Cell Biol. Int. Reports* **13**, 23–35.

Hauxwell, A.J., Corke, F.M.K., Hedley, C.L. and Wang, T.L. (1990) Storage protein gene expression is localised to regions lacking mitotic activity in developing pea embryos. An analysis of seed development in *Pisum sativum* L. XIV. *Development* **110**, 283–289.

Hauxwell, A.J., Corke, F.M.K. and Wang, T.L. (1993) Temporal and spatial gene expression in relation to cell division in the pea embryo. In: *Oscillations and Morphogenesis* (ed. L. Rensing). Marcel Dekker, New York, pp. 249–258.

Heath, J.D., Weldon, R., Monnot, C. and Meinke, D.W. (1986) Analysis of storage proteins in normal and aborted seeds from embryo-lethal mutants of *Arabidopsis thaliana*. *Planta* **169**, 304–312.

Johansen, D.A. (1950) *Plant Embryology: Embryogeny of the Spermatophyta*. Chronica Botanica Company, Waltham, MA.

Johnson, S., Liu, C.-M., Hedley, C.L. and Wang, T.L. (1994) An analysis of seed development in *Pisum sativum*. XVIII. The isolation of mutants defective in embryo development. *J. Exp. Bot.* **45**, 1503–1511.

Johri, B.M. (1984). *Embryology of Angiosperms*. Springer-Verlag, Berlin, Heidelberg.

Johri, B.M., Ambegaokar, K.B. and Srivastava, P.S. (1992) *Comparative Embryology of Angiosperms*. Springer-Verlag, Berlin, Heidelberg.

Karssen, C.M. and van Loon, L.C. (1992) Probing hormone action in developing seeds by ABA-deficient and insensitive mutants. In: *Progress in Plant Growth Regulation* (eds C.M. Karssen, C.L. van Loon and D. Vreugdenhil). Kluwer, Dordrecht, pp.43–53.

Lersten, N.R. (1983) Suspensors in Leguminosae. *Bot. Rev.* **49**, 233–257.

Liu, C.-M., Xu, Z.H., and Chua, N.-H. (1993) Auxin polar transport is essential for the establishment of bilateral symmetry during early plant embryogenesis. *Plant Cell* **5**, 621–630.

Liu, C-M., Johnson, S. and Wang, T.L. (1995) *cyd*, a mutant of pea that alters embryo morphology is defective in cytokinesis. *Dev. Genet.* **16**, 321–331.

Maheshwari, P. (1950) *An Introduction to the Embryology of the Angiosperms*. McGraw-Hill, New York.

Marinos, N.G. (1970a) Embryogenesis of the pea (*Pisum sativum*). I. The cytological environment of the developing embryo. *Protoplasma* **70**, 261–279.

Marinos, N.G. (1970b) Embryogenesis of the pea (*Pisum sativum*). II. An unusual type of plastid in the suspensor cells. *Protoplasma* **71**, 227–233.

Mayer, U., Torres-Ruiz, R.A., Berleth, T., Miséra, S. and Jürgens, G. (1991) Mutations affecting body organization in the *Arabidopsis* embryo. *Nature* **353**, 402–407.

Mendel, G. (1865) Versuche über pflanzen-hybriden. *Verh. Naturforsch. Ver. Brünn* **4**, 3–47.

Murfet, I.C. and Reid, J.B. (1993). Developmental mutants. In: *Peas: Genetics, Molecular*

Biology and Biotechnology (eds R. Casey and D.R. Davies). CAB International, Wallingford, pp. 165–216.

Nagl, W. (1990) Translocation of putrescine in the ovule, suspensor and embryo of *Phaseolus coccineus. J. Plant Physiol.* **136**, 587–591.

Patton, D.A. and Meinke, D.W. (1990) Ultrastructure of arrested embryos from lethal mutants of *Arabidopsis thaliana. Am. J. Bot.* **77**, 653–661.

Perez-Grau, L. and Goldberg, R.B. (1989) Soybean seed proteins are regulated spatially during embryogenesis. *Plant Cell* **1**, 1095–1109.

Prakash, N. (1987) Embryology of the Leguminosae. In: *Advances in Legume Systematics, Part 3* (ed. C.H. Stirton). Royal Botanic Gardens, Kew, pp. 241–278.

Reeve, R.M. (1948) Late embryogeny and histogenesis in *Pisum. Am. J. Bot.* **35**, 591–602.

Scharpé, A. and Van Parijs, R. (1973) The formation of polyploid cells in ripening cotyledons of *Pisum sativum* L. in relation to ribosome and protein synthesis. *J. Exp. Bot.* **24**, 216–222.

Schroeder, H.E., Schotz, A.H., Wardley-Richardson, T., Spencer, D. and Higgins, T.J.V. (1993) Transformation and regeneration of two cultivars of pea (*Pisum sativum* L.). *Plant Physiol.* **101**, 751–757.

Schwartz, B.W., Yeung, E.C. and Meinke, D.W. (1994) Disruption of morphogenesis and transformation of the suspensor in abnormal *suspensor* mutants of *Arabidopsis. Development* **120**, 3235–3245.

Smith, D.L. (1973) Nucleic acid, protein, and starch synthesis in developing cotyledons of *Pisum arvense* L. *Ann. Bot.* **37**, 795–804.

Smith, D.L. (1974) A histological and histochemical study of the cotyledons of *Phaseolus vulgaris* during germination. *Protoplasma* **79**, 41–57.

Smith, D.L. (1981) Cotyledons of the Leguminosae. In: *Advances in Legume Systematics, Part 2* (eds R.M. Polhill and P.H. Raven). Royal Botanic Gardens, Kew, pp. 927–940.

Smith, D.L. and Flinn, A.M. (1967) Histology and histochemistry of the cotyledons of *Pisum arvense* L. during germination. *Planta* **74**, 75–85.

Souèges, R. (1948) Embryogénie des Papilionacés. Développement de l'embryon chez le *Pisum sativum. Compt. Rend. Hebd. Séances Acad. Sci.* **227**, 802–804.

Wang, T.L. and Hedley, C.L. (1991) Seed development in peas: knowing your three 'r's' (or four, or five). *Seed Sci. Res.* **1**, 3–14.

Wang, T.L. and Hedley, C.L. (1993) Genetic and developmental analysis of the seed. In: *Peas: Genetics, Molecular Biology and Biotechnology* (eds R. Casey and D.R. Davies). CAB International, Wallingford, pp. 83–120.

Wang, T.L., Hadavizideh, A., Harwood, A., Welham, T.J., Harwood, W.A., Faulks, R. and Hedley, C.L. (1990) An analysis of seed development in *Pisum sativum*. XIII. The chemical induction of storage product mutants. *Plant Breeding* **105**, 311–320.

Yadegari, R., de Paiva, G., Laux, T., Koltunow, A.M., Apuya, N., Zimmerman, J.L., Fischer, R.L., Harada, J.L. and Goldberg, R.B. (1994) Cell differentiation and morphogenesis are uncoupled in *Arabidopsis raspberry* embryos. *Plant Cell* **6**, 1713–1729.

Yang, L.-J., Barratt, D.H.P., Domoney, C., Hedley, C.L. and Wang, T.L. (1990) An analysis of seed development in *Pisum sativum* L. Expression of storage protein genes in cultured embryos. *J. Exp. Bot.* **41**, 283–288.

Yeung, E.C. (1980) Embryogeny of *Phaseolus*: the role of the suspensor. *Z. Pflanzenphysiol.* **96**, 17–28.

Yeung, E.C. and Clutter, M.E. (1979) Embryogeny of *Phaseolus coccineus:* the ultrastructure and development of the suspensor. *Can. J. Bot.* **57**, 120–136.

Index